CLIMATE CHAN
AND THE REFUGI

Climate Change, Disasters, and the Refugee Convention is concerned with refugee status determination (RSD) in the context of disasters and climate change. It demonstrates that the legal predicament of people who seek refugee status in this connection has been inconsistently addressed by judicial bodies in leading refugee law jurisdictions, and identifies epistemological as well as doctrinal impediments to a clear and principled application of international refugee law. Arguing that RSD cannot safely be performed without a clear understanding of the relationship between natural hazards and human agency, the book draws insights from disaster anthropology and political ecology that see discrimination as a contributory cause of people's differential exposure and vulnerability to disaster-related harm. This theoretical framework, combined with insights derived from the review of existing doctrinal and judicial approaches, prompts a critical revision of the dominant human rights-based approach to the refugee definition.

MATTHEW SCOTT heads the People on the Move thematic area at the Raoul Wallenberg Institute of Human Rights and Humanitarian Law in Sweden, where he leads research and educational initiatives relating to human rights, disasters and displacement in Asia and the Pacific, Africa, and Europe. He is a solicitor of England and Wales and practised asylum and immigration law before entering academia.

CAMBRIDGE ASYLUM AND MIGRATION STUDIES

At no time in modern history have so many people been on the move as at present. Migration facilitates critical social, economic, and humanitarian linkages. But it may also challenge prevailing notions of bounded political communities, of security, and of international law.

The political and legal systems that regulate the transborder movement of persons were largely devised in the mid-twentieth century, and are showing their strains. New challenges have arisen for policy-makers, advocates, and decision-makers that require the adaptation and evolution of traditional models to meet emerging imperatives.

Edited by a world leader in refugee law, this new series aims to be a forum for innovative writing on all aspects of the transnational movement of people. It publishes single or co-authored works that may be legal, political, or cross-disciplinary in nature, and will be essential reading for anyone looking to understand one of the most important issues of the twenty-first century.

Series Editor
James Hathaway, James E. and Sarah A. Degan Professor of Law, and Director of Michigan Law's Program in Refugee and Asylum Law, University of Michigan, USA

Editorial Advisory Board
Alexander Betts, Leopold Muller Professor of Forced Migration and International Affairs, and the Director of the Refugee Studies Centre, University of Oxford, UK
Vincent Chetail, Professor of Public International Law, and Director of the Global Migration Centre, Graduate Institute of International and Development Studies, Switzerland
Thomas Gammeltoft-Hansen, Professor with Special Responsibilities in Migration and Refugee Law at the University of Copenhagen
Audrey Macklin, Professor and Chair in Human Rights Law, University of Toronto, Canada
Saskia Sassen, Robert S. Lynd Professor of Sociology, and Chair of the Committee on Global Thought, Columbia University, USA

Books in the Series
The Child in International Refugee Law Jason Pobjoy
Refuge Lost: Asylum Law in an Interdependent World Daniel Ghezelbash
Demanding Rights: Europe's Supranational Courts and the Dilemma of Migrant Vulnerability Moritz Baumgärtel
Climate Change, Disasters and the Refugee Convention Matthew Scott

CLIMATE CHANGE, DISASTERS, AND THE REFUGEE CONVENTION

MATTHEW SCOTT

Raoul Wallenberg Institute of Human Rights and Humanitarian Law

CAMBRIDGE
UNIVERSITY PRESS

University Printing House, Cambridge CB2 8BS, United Kingdom

One Liberty Plaza, 20th Floor, New York, NY 10006, USA

477 Williamstown Road, Port Melbourne, VIC 3207, Australia

314–321, 3rd Floor, Plot 3, Splendor Forum, Jasola District Centre, New Delhi – 110025, India

79 Anson Road, #06–04/06, Singapore 079906

Cambridge University Press is part of the University of Cambridge.

It furthers the University's mission by disseminating knowledge in the pursuit of education, learning, and research at the highest international levels of excellence.

www.cambridge.org
Information on this title: www.cambridge.org/9781108478229
DOI: 10.1017/9781108784580

© Matthew Scott 2020

This publication is in copyright. Subject to statutory exception and to the provisions of relevant collective licensing agreements, no reproduction of any part may take place without the written permission of Cambridge University Press.

First published 2020

Printed in the United Kingdonm by TJ International Ltd, Padstow, Cornwall

A catalogue record for this publication is available from the British Library.

Library of Congress Cataloging-in-Publication Data
Names: Scott, Matthew, 1978– author.
Title: Climate change, disasters, and the refugee convention / Matthew Scott, Raoul Wallenberg Institute of Human Rights and Humanitarian Law.
Description: Cambridge, United Kingdom ; New York, NY, USA : Cambridge University Press, 2020. | Series: Cambridge asylum and migration studies | Based on author's thesis (doctoral - Lunds universitet Juridiska fakulteten, 2018) issued under title: Refugee status determination in the context of 'natural' disasters and climate change : a human rights-based approach. | Includes bibliographical references and index.
Identifiers: LCCN 2019042228 (print) | LCCN 2019042229 (ebook) | ISBN 9781108478229 (hardback) | ISBN 9781108747127 (paperback) | ISBN 9781108784580 (epub)
Subjects: LCSH: Refugees–Legal status, laws, etc. | Asylum, Right of. | Convention Relating to the Status of Refugees (1951 July 28). | Environmental refugees–Legal status, laws, etc. | Disaster victms–Legal status, laws, etc.
Classification: LCC KZ6530 .S36 2020 (print) | LCC KZ6530 (ebook) | DDC 341.4/86–dc23
LC record available at https://lccn.loc.gov/2019042228
LC ebook record available at https://lccn.loc.gov/2019042229

ISBN 978-1-108-47822-9 Hardback
ISBN 978-1-108-74712-7 Paperback

Cambridge University Press has no responsibility for the persistence or accuracy of URLs for external or third-party internet websites referred to in this publication and does not guarantee that any content on such websites is, or will remain, accurate or appropriate.

CONTENTS

List of Figures — page viii
Series Editor's Preface — ix
Acknowledgements — x
Note on the Text — xii
Table of Cases — xiii
Table of Treaties and Other International and Regional Instruments — xix
List of Abbreviations — xxi

1 Introduction — 1
   1.1 Introduction — 1
   1.2 The Dominant View — 1
   1.3 A Different Perspective — 6
   1.4 Structure of the Book — 8

2 Two Disaster Paradigms — 10
   2.1 What Is a Disaster? — 10
   2.2 The Hazard Paradigm — 12
   2.3 The Social Paradigm — 15
      2.3.1 Vulnerability — 16
      2.3.2 The Pressure and Release Model — 18
      2.3.3 Definition — 21
      2.3.4 Discrimination — 22
      2.3.5 Structural Violence — 28

3	**Jurisprudence on the Determination of Refugee Status in the Context of 'Natural' Disasters and Climate Change**	**32**
	3.1 Methodology	32
	3.2 Basic Principles of International Refugee Law	33
	3.2.1 Well-Founded Fear	34
	3.2.2 ... of Being Persecuted	35
	3.2.3 ... for Reasons of	37
	3.2.4 ... Race, Religion, Nationality, Membership of a Particular Social Group, or Political Opinion	44
	3.2.5 Internal Relocation	44
	3.3 Structure of the Review	45
	3.4 Disasters and Climate Change Are Peripheral to the Majority of Claims in the Review	45
	3.5 Category 1: Indiscriminate Adversity due to the Forces of Nature	48
	3.6 Category 2: Direct and Intentional Infliction of Harm	54
	3.6.1 The State Intentionally Causing Environmental Damage in Order to Harm a Particular Group	54
	3.6.2 Crackdowns on (Perceived) Dissent Relating to the Causes and/or Management of Environmental Degradation or Disasters	55
	3.6.3 Discriminatory Denial of Disaster Relief	57
	3.7 Category 3: Other Failures of State Protection	63
	3.7.1 *Ex post* Failures of State Protection	63
	3.7.2 *Ex ante* Failures of Protection	68
	3.7.3 Combined *Ex post* and *Ex ante* Claims	69
	3.8 Unresolved Doctrinal Issues Arising from the Review	87
4	**Interpreting the Refugee Definition**	**89**
	4.1 The Unitary Character of the Refugee Definition	89
	4.2 The Relevance of International Human Rights Law to the Interpretation of Being Persecuted	90
	4.3 The Methodology Prescribed by the VCLT	93
5	**The Temporal Scope of Being Persecuted**	**96**
	5.1 Being Persecuted from the Perspective of the 'Event Paradigm'	96

	5.2 The 'Well-Founded Fear' Element as Risk Assessment	100
	5.3 Why a Predicament Approach to Determining the Existence of a Well-Founded Fear of Being Persecuted Is Appropriate	107

6 The Personal Scope of Being Persecuted: The Function of the Non-discrimination Norm within the Refugee Definition — 112

6.1 The Non-discrimination Norm — 112

6.2 The Non-discrimination Norm in International Refugee Law — 115

    6.2.1 The Non-discrimination Norm Reflected in the Nexus Clause — 117

    6.2.2 The Non-discrimination Norm Reflected in the Notion of Being Persecuted — 118

6.3 Discrimination as an Inherent Feature of Being Persecuted — 121

    6.3.1 Challenges to the Proposition that Discrimination Is an Inherent Feature of Being Persecuted — 125

6.4 Definition of Being Persecuted — 128

7 Refugee Status Determination in the Context of 'Natural' Disasters and Climate Change — 132

7.1 The Test — 132

7.2 Step One: Being Persecuted for a Convention Reason — 132

    7.2.1 The First Element: Discrimination for a Convention Reason — 133

    7.2.2 The Second Element: (The Risk of) Exposure to Serious Denials of Human Rights — 133

    7.2.3 The Third Element: Nexus — 143

7.3 Step Two: The Standard of Proof — 148

7.4 Step Three: Internal Relocation — 149

7.5 Circumstances Not Reflecting a Well-Founded Fear of Being Persecuted for a Convention Reason — 150

7.6 Conclusion — 155

Appendix: Taxonomy — 157

Bibliography — 159

Index — 168

FIGURES

7.1 Somalia IPC July 2011   *page* 137
7.2 Somalia IPC June 2017   138
7.3 Local Risk Index: case study Indonesia   141
7.4 Acute food insecurity: near term (February–May 2017)   143
7.5 Acute food insecurity: medium term (June–September 2017)   144

SERIES EDITOR'S PREFACE

Two of the most pressing challenges facing the world today are climate change and involuntary migration. Not only is each a real concern in its own right, but it is also increasingly clear that the two phenomena are deeply connected: climate change and other 'natural' disasters are not only devastating the environment, but they are also forcing more and more people to abandon their homes. While most climate-induced migration has thus far occurred within states, a growing number of persons feels compelled to seek protection abroad. And as the scale and pace of climate-induced migration increase, it is clear that more and more people will seek protection abroad as refugees. Is refugee status the right remedy for persons fleeing climate change and comparable disasters?

In this pioneering study, Matthew Scott begins by challenging the traditional objections to the recognition of refugee status in such circumstances – that the requisite element of discrimination is lacking, and that the harms involved are not 'persecutory'. But he goes much farther, engaging deeply and thoughtfully with the underlying 'hazard paradigm' that informs dominant understandings of refugee status as a whole. He makes a persuasive case that we can only do justice to emerging protection claims by paying more attention to the discriminatory social contexts within which exposure to serious harm arises. By making this shift to a 'social paradigm' of refugee status we can, he argues, see in the vulnerability of many individuals and groups precisely the hallmarks that should be understood to be the basis for the recognition of refugee status at international law.

Scott's analysis is especially important because it is both conceptually provocative and deeply attentive to relevant jurisprudence from around the world. This is no glib recipe for a reform that might be as unrealistic as it is appealing. *Climate Change, Disasters, and the Refugee Convention* instead beautifully bridges the divide between respect for extant law and legal process on the one hand, and the importance of pursuing reinvention in the service of critical human needs on the other. The result is a book that not only presents us with the first comprehensive analysis of the challenges of invoking refugee law to respond to the phenomenon of climate-induced migration, but that also leverages that analysis to push the boundaries of legal understandings of refugee protection in new and potentially productive directions.

James C. Hathaway
Editor, Cambridge Asylum and Migration Studies

ACKNOWLEDGEMENTS

This book is the product of six years' worth of discussions with colleagues at Lund University and wider refugee law academic and practitioner communities.

First, I want to thank Gregor Noll and Rebecca Stern for their supervision. I also thank Reza Banakar for his close engagement with the research in the early stages.

I thank Mark Gibney and Karol Nowak for their thoughtful and critical engagement with my text when it was beginning to mature, and Boldizsár Nagy for his careful critique of the mature draft.

I want to thank the many academic colleagues who have engaged in discussion or taken time to scrutinize academic writing that has informed the development of this book, including Jane McAdam, James Hathaway, David Cantor, Geoff Gilbert, Elspeth Guild, Walter Kälin, Ben Saul, Ulf Linderfalk, Thomas Gammeltoft-Hansen, and Morten Kjærum.

I have also benefitted greatly from insightful discussions with practitioners, including Nuala Mole, Frances Webber, Satvinder Juss, Hugo Storey, Julian Bild, Jawaid Luqmani, Judith Gleeson, Simon Cox, Madelaine Seidlitz, Bruce Burson, Sheona York, Alison Harvey, and Carl Söderbergh.

I am particularly indebted to Kalvir Kaur, who inducted me into the practice of refugee law, and to the many other creative, passionate immigration and refugee law practitioners I had the honour and privilege of working with.

I am very grateful to the welcome extended to me by the Kaldor Centre for International Refugee Law at the University of New South Wales, recalling fondly the stimulating intellectual atmosphere just minutes from the beach. I would like to mention in particular Frances Voon, Claire Higgins, Madeline Gleeson, Luke Potter, Khanh Hoang, Vicky Sentas, as well as Rosemary Rayfuse for introducing me to colleagues within the wider law faculty. Financial support from the Knut and Alice Wallenbergs Foundation and the Landshövding Per Westlings Memorial Fund is gratefully acknowledged.

Financial support by the European Refugee Fund for the first part of my doctoral studies is gratefully acknowledged.

I would like to recall the stimulating academic and social space created by the Lund/Uppsala Migration Law Network (LUMIN), enlivened by Aleksandra Popovic, Anna Bruce, Lisa Kerker, Martin Ratkovich, Isa Cegrell Carlander, Maria Bexelius, Hanna Wikström, and Emily Diab, and the guiding vision provided by Gregor Noll and Rebecca Stern. I have worked particularly closely with Vladislava Stoyanova and Eleni Karageorgiou, also LUMIN founding members, and have thoroughly enjoyed their friendship, collegiality, and critical engagement with my text over time.

I would also like to extend sincere appreciation to my opponent at the doctoral defence, Michelle Foster, as well as members of the examination committee, including Lena Halldenius, Mo Hamza, and Jens Vedsted-Hansen.

I also express my great pleasure at having the opportunity to work now on issues related to those arising in this book with the very many fine colleagues at the Raoul Wallenberg Institute of Human Rights and Humanitarian Law.

Finally, for countless reasons, I thank Hanna Scott, and our kids Silas and Caspian, to whom this book is dedicated.

NOTE ON THE TEXT

In the absence of a neutral third person singular pronoun in the English language, this book adopts the female pronouns 'she' and 'her', rather than opting for the clumsy 'he or she' and 'his or her'.

Owing to the multiple terms applied to individuals at different stages in the process of seeking recognition of refugee status, the term 'claimant' is used throughout the book as a general term. Where appropriate, more specific terms such as 'appellant' or 'applicant' are used.

TABLE OF CASES

International Court of Justice

Avena and Other Mexican Nationals (Mexico v United States) [2004] ICJ Rep 12, 93
Barcelona Traction, Light and Power Company, Limited (Judgment) [1970] ICJ Rep 3, 112
Legal Consequences for States of the Continued Presence of South Africa in Namibia (South West Africa) notwithstanding Security Council Resolution 276 (1970) (Advisory Opinion) [1971] ICJ Rep 16, 112

European Court of Human Rights

AL v Austria (2012) ECHR 828, 56, 157
Budayeva and Others v Russian Federation (2014) 59 EHRR 2, 69, 79, 87
Cruz Varas and Others v Sweden (1991) 14 EHRR 1, 101
DH and Others v Czech Republic (2008) 47 EHRR 3, 114
Golder v United Kingdom (1979) 1 EHRR 524Öneryildiz v Turkey (2005) 41 EHRR 20, 93
Soering v United Kingdom (1989) 11 EHRR 439, 91
TN and SN v Denmark (2011) ECHR 94, 47

Inter-American Court of Human Rights

Advisory Opinion OC-18/03 of 17 September 2003, Juridical Condition and Rights of Undocumented Migrants, 112
Advisory Opinion OC-21/14 of August 19, 2014 requested by the Argentine Republic, the Federative Republic of Brazil, the Republic of Paraguay and the Oriental Republic of Uruguay. Rights and Guarantees of Children in the Context of Migration and/or In Need of International Protection, OC-21/14, Inter-American Court of Human Rights (IACrtHR), 19 August 2014, 33
PM 340/10 – Women and girls residing in 22 Camps for internally displaced persons in Port-au-Prince, Haiti, 135

United Nations Human Rights Monitoring Bodies

Aaldersberg and Others *v* Netherlands CCPR/C /87/D/1440/2005 (14 August 2006), 79, 82, 134
Simunek et al *v* Czech Republic (HRC 516/92), 114
Zwaan de Vries *v* the Netherlands CCPR/C/29/D/182/1984, 114

Australia

Applicant A *v* Minister for Immigration and Ethnic Affairs [1997] HCA 4, (1997) 190 CLR 225, 2, 3, 33, 52, 83, 89, 91, 94, 119
Chan *v* Minister for Immigration and Ethnic Affairs [1989] HCA 62, (1989) 169 CLR 379, 34, 100, 129, 141
Chen Shi Hai *v* Minister for Immigration and Multicultural Affairs [2000] HCA 19, (2000) 201 CLR 293, 39, 41
Hagi-Mohammed *v* Minister for Immigration and Multicultural Affairs [2001] FCA 1156, 58, 157
Minister for Immigration *v* Haji Ibrahim [2000] HCA 55, (2000) 204 CLR 1, 98, 120
Minister for Immigration and Multicultural Affairs *v* Khawar [2002] HCA 14, (2002) 210 CLR 1, 122
Minister for Immigration and Multicultural Affairs *v* Respondents S152/2003 (2004) 222 CLR 1, 92
Mohamad *v* Minister for Immigration and Multicultural Affairs [1999] FCA 688, 66
Mohammed Motahir Ali *v* Minister of Immigration, Local Government and Ethnic Affairs [1994] FCA 887, 48
NACM of 2002 *v* Minister for Immigration and Multicultural and Indigenous Affairs [2003] FCA 1554, 40, 41, 128
NAGT of 2002 *v* Minister for Immigration and Multicultural and Indigenous Affairs [2002] FCAFC 319, 105
Perampalam *v* Minister for Immigration and Multicultural Affairs [1999] FCA 165, (1999) 84 FCR 274, 149
SZOGB *v* Minister for Immigration and Another [2010] FMCA 748, 33
Ram *v* Minister for Immigration and Ethnic Affairs, (1995) 130 ALR 213 (Aus. FFC, 27 June 1995), 39, 120
RRT Case No. N95/09386 [1996] RRTA 3191 (7 November 1996), 50
RRT Case No. N96/10806 [1996] RRTA 3195 (7 November 1996), 50
RRT Case No. N93/00894 [1996] RRTA 3244 (14 November 1996), 68, 157
RRT Case No. N99/30231 [2000] RRTA 17 (10 January 2000), 52
RRT Case No. 060793741 [2007] RRTA 3 (11 January 2007), 46
RRT Case No. 061051351 [2007] RRTA 15 (2 February 2007), 46
RRT Case No. 060926579 [2007] RRTA 36 (28 February 2007), 57, 157
RRT Case No. 071204406 [2007] RRTA 86 (27 April 2007), 46
RRT Case No. 071295385 [2007] RRTA 109 (20 June 2007), 65, 67, 158
RRT Case No. 0901657 [2009] RRTA 526 (11 June 2009), 49
RRT Case No. 0903027 [2009] RRTA 1102 (28 October 2009), 46

RRT Case No. 0907346 [2009] RRTA 1168 (10 December 2009),   51, 52
RRT Case No. 0903555 [2010] RRTA 31 (15 January 2010),   56, 157
RRT Case No. 1001325 [2010] RRTA 373 (11 May 2010),   57, 157
RRT Case No. 1002650 [2010] RRTA 595 (14 July 2010),   60
RRT Case No. 1004726 [2010] RRTA 845 (30 September 2010),   52
RRT Case No. 1002915 [2011] RRTA 615 (14 July 2011),   47
RRT Case No. 1104064 [2011] RRTA 659 (3 August 2011),   57–157
RRT Case No. 1200203 [2012] RRTA 145 (6 March 2012),   69–72, 158
RRT Case No. 1203764 [2012] RRTA 312 (18 May 2012),   46
RRT Case No. 1211848 [2012] RRTA 933 (16 October 2012),   47
RRT Case No. 1209786 [2012] RRTA 1058 (29 November 2012),   47
RRT Case No. 1215708 [2012] RRTA 1084 (6 December 2012),   47
RRT Case No. 1203936 [2012] RRTA 1070 (14 December 2012),   47
RRT Case No. 1213171 [2013] RRTA 157 (15 February 2013),   33

Canada

Alfarsy v Canada (Minister of Citizenship and Immigration) [2003] FCJ No. 1856; 2003 FC 1461,   33
Canada (Attorney General) v Ward [1993] 2 SCR 689,   2, 83, 94
Chan v Canada [1993] 42 ACWS 3d 259,   58, 157
Ferguson v Canada (Minister of Citizenship and Immigration) 2008 FC 903, 169 ACWS (3d) 629,   63, 158
Gladys Maribel Hernandez, Immigration Appeal Board Decision M81-1212, January 6, 1983,   97
Pushpanathan v Canada (Minister of Citizenship and Immigration) [1998] 1 SCR 982, 1024,   91

Germany

Bundesverwaltungsgericht, Beschluss vom 7. Februar 2008 – BVerwG 10 C 33.07,   2

Ireland

S.J.L. v Refugee Appeals Tribunal and Others [2014] IEHC 608,   2

New Zealand

AC (Tuvalu) [2014] NZIPT 800517-520,   74–87, 134, 153, 155, 158
AD (Tuvalu) [2014] NZIPT 501370-371,   85

AF (Kiribati) [2013] NZIPT 800413,   54, 74–87, 105, 134, 157, 158
AF (Tuvalu) [2015] NZIPT 800859,   85
AV (Nepal) [2017] NZIPT 801125–126,   85
BG (Fiji) [2012] NZIPT 800091,   76, 81, 94
CV v Immigration and Protection Tribunal and CW v Immigration and Protection Tribunal [2015] NZHC 51,   123, 124
DS (Iran) [2016] NZIPT 800788 (2 February 2016),   37, 100, 108, 123–125, 130
K v Refugee Status Appeals Authority (HC Auckland, M 1586-SW99, February 2000),   49
Refugee Appeal No. 11/91 RSAA (5 September 1991),   76
Refugee Appeal No. 70074/96 RSAA (17 September 1996),   34, 117
Refugee Appeal No. 70959/98 RSAA (27 August 1998),   65
Refugee Appeal No. 70965/98 RSAA 289 (27 August 1998),   64, 158
Refugee Appeal No. 71427/99 RSAA (16 August 2000),   37, 122
Refugee Appeal No. 72186/2000 RSAA (10 August 2000),   53
Refugee Appeal Nos. 72189–72195/2000 (17 August 2000),   53
Refugee Appeal No. 72635/01 RSAA (6 September 2002),   40, 41, 43
Refugee Appeal No. 74665/03 RSAA (7 July 2004),   76, 89, 90, 107, 110
Refugee Appeal No. 75655 RSAA (29 September 2006),   46
Refugee Appeal No. 76074 RSAA (22 November 2007),   46
Refugee Appeal No. 76191 RSAA (12 August 2008),   46
Refugee Appeal No. 76237 RSAA (15 December 2008),   60, 157
Refugee Appeal No. 76374 RSAA (28 October 2009),   55–56, 77, 157
Refugee Appeal No. 76457 RSAA (15 March 2010),   46
S v Chief Executive of the Department of Labour and Anor [2005] NZHC 439, HC AK CIV 2005-404-003360,   34, 49
Teitiota v Chief Executive of the Ministry of Business Innovation and Employment [2013] NZHC 3125,   82
Teitiota v The Chief Executive of the Ministry of Business Innovation and Employment [2014] NZCA 173,   83
Teitiota v The Chief Executive of the Ministry of Business, Innovation and Employment [2015] NZSC 107,   84

Norway

Borgarting Court of Appeal Decision of 23 September 2011. Abid Hassan Jama v Utlendingsnemnda, 10-142363ASD-BORG/01, Norway,   46

United Kingdom

14 (Kabul – Pashtun) Afghanistan v Secretary of State for the Home Department, CG [2002] UKIAT 05345,   46
AK (Article 15(c)) Afghanistan v Secretary of State for the Home Department, CG [2012] UKUT,   46

TABLE OF CASES

AN (Tunni Torre) [2004] UKIAT 270 (24 September 2004), 146
DJ (Bantu – Not Generally at Risk) Somalia v Secretary of State for the Home Department CG [2005] UKIAT 00089 (22 April 2005), 146, 147
Doymus v Secretary of State for the Home Department (IAT, HX/80112/99, 19 July 2000), 98, 99
GS (Article 15(c) – Indiscriminate Violence) Afghanistan v Secretary of State for the Home Department CG [2009] UKAIT 00044, 46
HH and Others (Mogadishu – Armed Conflict – Risk) Somalia v Secretary of State for the Home Department, CG [2008] UKAIT 00022, 46
HJ (Iran) v Secretary of State for the Home Department [2010] UKSC 31, [2011] AC 596, 106
HK and Others Afghanistan v Secretary of State for the Home Department CG [2010] UKUT 378, 46
HM and Others (Article 15(c)) Iraq CG v the Secretary of State for the Home Department, [2012] UKUT 00409, 46
Horvath v Secretary of State for the Home Department [2000] UKHL 37, [2001] AC 489, 2, 4, 5, 37, 83, 94, 98, 163
HS (returning asylum seekers) Zimbabwe CG [2007] UKAIT 00094, 59, 157
Islam v Secretary of State for the Home Department Immigration Appeal Tribunal and Another, ex parte Shah, R v [1999] UKHL 20, [1999] AC 629, 36, 98, 120
Januzi v Secretary of State for the Home Department [2006] UKHL 5, [2006] 2 AC 426, 44
Karanakaran v Secretary of State for the Home Department [2000] EWCA Civ. 11, [2000] Imm AR 271, 33, 89
LM (Relocation – Khartoum – AE Reaffirmed) Sudan v Secretary of State for the Home Department [2005] UKAIT 00114, 46
MA (Somalia) v Secretary of State for the Home Department [2009] EWCA Civ 4, 46
Omoruyi [2001] Imm AR 175, 128
R v Governor of Pentonville Prison, ex parte Fernandez [1971] 2 All ER 691, [1971] 1 WLR 987, 104
R v IAT ex parte Shah [1997] Imm AR 145, 149
R v Secretary of State for the Home Department, ex parte Bugdaycay [1987] AC 514, 49
R v Secretary of State for the Home Department, ex parte Ahmed Aissaoui United Kingdom High Court (England and Wales) (11 November 1996) CO/1391/96, 47
R v Secretary of State for the Home Department, ex parte Sivakumaran and Conjoined Appeals (UN High Commissioner for Refugees Intervening) [1987] UKHL 1, [1988] AC 958, 34, 94, 100, 103, 104, 129, 141
RN (Returnees) Zimbabwe CG [2008] UKAIT 00083, 58, 59, 60, 62, 157
RS and Others (Zimbabwe – AIDS) Zimbabwe v Secretary of State for the Home Department CG [2010] UKUT 363, 60, 61, 157
Sepet and Bulbul [2003] 3 All ER 304, [2003] 1 WLR 856, 128
SH (Rahanweyn not a minority clan) Somalia CG [2004] UKIAT 00272 (28 September 2004), 146
Ullah v Special Adjudicator [2004] UKHL 26, 61

United States of America

Chacon v INS, 341 F.3d 533, United States Court of Appeals for the Sixth Circuit, 18 August 2003, 46

In re Katrina Canal Breaches Consolidated Litigation (Robinson) 647 F. Supp. 2d 644 (E.D. La. 2009), 151, 152

INS v Cardoza-Fonseca 480 US 421 (1987), 34, 103, 104, 129, 141

Immigration and Naturalization Service v Elias Zacarias, (1992) 502 US 478 (USSC, 22 January 1992), 38

Matter of Acosta, A-24159781, United States Board of Immigration Appeals, 1 March 1985, 2, 96

Matter of Echeverria, 25 I&N Dec. 512 (BIA 2011), United States Board of Immigration Appeals, 1 June 2011, 32

Matter of Maria Armida Sosa Ventura, 25 I&N Dec. 391 (BIA 2010), United States Board of Immigration Appeals, 23 November 2010, 2

Milat v Holder United States Court of Appeals, Fifth Circuit June 19, 2014 755 F.3d 354, 101

YC v Holder, Attorney General, 11-2749-ag, 11-3217-ag, United States Court of Appeals for the Second Circuit (18 December 2013), 57, 157

TABLE OF TREATIES AND OTHER INTERNATIONAL AND REGIONAL INSTRUMENTS

1933　League of Nations, Convention Relating to the International Status of Refugees (28 October 1933) League of Nations, Treaty Series Vol. CLIX No. 3663   101

1945　United Nations, Charter of the United Nations (24 October 1945) 1 UNTS XVI,   112

1946　United Nations, Constitution of the International Refugee Organization (15 December 1946) 18 UNTS 3,   101

1948　Universal Declaration of Human Rights (adopted 10 December 1948) UNGA Res 217 A(III),   36, 90, 94, 97, 112, 117, 119, 125, 126

1951　Convention Relating to the Status of Refugees (adopted 28 July 1951, entered into force 22 April 1954) 189 UNTS 137,   1, 33–44, 89–92

1965　International Convention on the Elimination of All Forms of Racial Discrimination (adopted 21 December 1965, entered into force 4 January 1969) 660 UNTS 195,   112

1966　International Covenant on Civil and Political Rights (adopted 16 December 1966, entered into force 23 March 1976) 999 UNTS 171,   32, 36, 76, 78, 80–84, 112, 113, 117, 125, 134

1966　International Covenant on Economic, Social and Cultural Rights (adopted 16 December 1966, entered into force 3 January 1976) 993 UNTS 3,   36, 62, 63, 69, 76, 112, 114, 117, 133

1967　Protocol Relating to the Status of Refugees (adopted 31 January 1967, entered into force 4 October 1967) 606 UNTS 267 (Protocol),   1

1969　Convention Governing the Specific Aspects of Refugee Problems in Africa (adopted 10 September 1969, entered into force 20 June 1974) 1001 UNTS 45,   148

1969　Vienna Convention on the Law of Treaties (adopted 23 May 1969, entered into force 27 January 1980) 1155 UNTS 331,   2, 90, 93, 95, 109, 113, 115, 121, 124, 129, 131, 155

1979　Convention on the Elimination of All Forms of Discrimination Against Women (adopted 18 December 1979, entered into force 3 September 1981) 1249 UNTS 13,   112–114

1989　Convention on the Rights of the Child (adopted 20 November 1989, entered into force 2 September 1990) 1577 UNTS 3,   36, 85, 112–114

1990　International Convention on the Protection of the Rights of All Migrant Workers and Members of their Families (adopted 18 December 1990, entered into force 1 July 2003) 2220 UNTS 3,   112

1992	UN Framework Convention on Climate Change (adopted 9 May 1992, entered into force 21 March 1994) 1771 UNTS 107, 10
2006	Convention on the Rights of Persons with Disabilities (adopted 13 December 2006, entered into force 3 May 2008) 2515 UNTS 3, 112
2006	International Convention for the Protection of All Persons from Enforced Disappearance (adopted 20 December 2006, entered into force 23 December 2010) 2716 UNTS 3, 112
2011	Council of the European Union, Directive 2011/95/EU of the European Parliament and of the Council of 13 December 2011 on Standards for the Qualification of Third-Country Nationals or Stateless Persons as Beneficiaries of International Protection, for a Uniform Status for Refugees or for Persons Eligible for Subsidiary Protection, and for the Content of the Protection Granted (Recast), 20 December 2011, OJ L. 337/9-337/26; 20.12.2011, 61, 96, 99, 100, 108, 118

ABBREVIATIONS

CEDAW	Convention on the Elimination of All Forms of Discrimination against Women 1979 / Committee on the Elimination of All Forms of Discrimination against Women
CERD	Convention on the Elimination of All Forms of Racial Discrimination 1965 / Committee on the Elimination of All Forms of Racial Discrimination
CESCR	Committee on Economic, Social and Cultural Rights
CRC	Convention on the Rights of the Child 1989 / Committee on the Rights of the Child
ECHR	European Convention on Human Rights and Fundamental Freedoms 1950
ExCom	Executive Committee of UN High Commissioner for Refugees
FEWS	Famine Early Warning System
HRC	Human Rights Committee
IARLJ	International Association of Refugee Law Judges
IASC	Inter-Agency Standing Committee
ICCPR	International Covenant on Civil and Political Rights 1966
ICESCR	International Covenant on Economic, Social and Cultural Rights 1966
IFRC	International Federation of Red Cross and Red Crescent Societies
IHRL	International human rights law
ILC	International Law Commission
IPC	Integrated Food Security Phase Classification
IPCC	Intergovernmental Panel on Climate Change
IRO	International Refugee Organization
NZHC	High Court of New Zealand
NZIPT	New Zealand Immigration and Protection Tribunal
OHCHR	Office of the High Commissioner for Human Rights
RRTA	Refugee Review Tribunal of Australia
RSAA	New Zealand Refugee Status Appeals Authority
RSD	Refugee status determination
UKAIT	United Kingdom Asylum and Immigration Tribunal
UNFCCC	UN Framework Convention on Climate Change 1992

UNGA	UN General Assembly
UNHCR	UN High Commissioner for Refugees
UNHRC	UN Human Rights Council
UNISDR/UNDRR	UN Office for Disaster Risk Reduction
UNOCHA	UN Office for the Coordination of Humanitarian Affairs
UNSC	UN Security Council
VCLT	Vienna Convention on the Law of Treaties 1969

1

Introduction

1.1 Introduction

This book is concerned with the determination of refugee status (RSD) in the context of 'natural' disasters and climate change. The use of the quotation marks highlights a critical, paradigmatic understanding that registers the disasters that unfold in the context of natural hazards as deeply social processes. This approach differs from the notion of 'natural disasters', which tends to reflect a hazard-centric understanding of the phenomenon. As will be developed over the course of this book, how disasters are understood plays a role in RSD in this context.

Central to this book is the challenge that it presents to the dominant view that the legal meaning of the phrase 'a well-founded fear of being persecuted for reasons of race, religion, nationality, membership of a particular social group or political opinion' contained in the refugee definition at Article 1A(2) of the 1951 Convention Relating to the Status of Refugees and its 1967 Protocol[1] (the Refugee Convention) makes the Refugee Convention an inherently peripheral instrument for addressing the legal status of persons displaced across borders in the context of disasters and climate change. Critically, in this book I do not attempt to expand the refugee definition or offer any sort of policy proposal that would secure international protection for a greater number of people. Rather, I investigate assumptions underpinning the pronounced dichotomy between refugees and persons facing adversity in the context of disasters and climate change. Assumptions are both epistemological, relating to the perceived nature of 'natural' disasters, as well as doctrinal, concerning the meaning of the constitutive elements of the refugee definition at Article 1A(2). The result is a recalibrated human-rights-based understanding of the refugee definition that can be applied as much to claims for recognition of refugee status in the context of disasters and climate change as to other sorts of refugee claim.

1.2 The Dominant View

What in this book I refer to as the 'dominant view' of how the Refugee Convention applies in the context of 'natural' disasters and climate change is reflected in key cases that have shaped the development of international refugee law, and the writing of legal doctrinal scholars that purport to apply the law to the phenomenon of cross-border displacement in

---

[1] Convention Relating to the Status of Refugees (adopted 28 July 1951, entered into force 22 April 1954) 189 UNTS 137 and Protocol Relating to the Status of Refugees (adopted 31 January 1967, entered into force 4 October 1967) 606 UNTS 267 (Protocol).

this connection. This 'view' is not monolithic, but key features emerge that highlight the inherent difficulties in establishing a well-founded fear of being persecuted in the context of 'natural' disasters and climate change.

The pronouncement in *Matter of Acosta* by the United States Board of Immigration Appeals reflects positions adopted by the High Court of Australia,[2] the Supreme Court of England and Wales,[3] and the Canadian Supreme Court[4] amongst others,[5] that 'large groups of persons displaced by civil strife and natural disasters ... simply do not qualify under the Protocol's limited definition of a "refugee"'.[6]

Dawson J's obiter dictum in *A and Another* reflects the view that disaster-related harm is a phenomenon that is quite distinct from 'persecution':

> By including in its operative provisions the requirement that a refugee fear persecution, the Convention limits its humanitarian scope and does not afford universal protection to asylum seekers. No matter how devastating may be the epidemic, natural disaster or famine, a person fleeing them is not a refugee within the terms of the Convention.[7]

Dawson J expresses this view in the course of reaching an interpretation of the 'particular social group' ground within the refugee definition. Recognising that Article 31 of the Vienna Convention on the Law of Treaties[8] (VCLT) requires reference to the context in which the term is used and the object and purpose of the treaty, he is drawn to consider what the Refugee Convention was designed to achieve. He recognises the humanitarian aim of the Convention from a reference in the preamble, but also notes the heavy burden refugees can place on host countries. From this he draws the conclusion above that the requirement that a refugee fear persecution limits the humanitarian scope of the Convention. What 'persecution' means is therefore crucial to understanding the scope of the refugee definition. Whatever its meaning, it would appear from the above that it does not include victims of epidemics, natural disasters, or famines.

Unfortunately, Dawson J does not dwell on disasters in the remainder of his judgment, which is more concerned with determining the meaning of the membership of a particular social group ground. However, McHugh J, in the same case, assists with his clarification of the meaning of 'persecution for a Convention reason' under the Refugee Convention:

> Persecution for a Convention reason may take an infinite variety of forms from death or torture to the deprivation of opportunities to compete on equal terms with other members of the relevant society. Whether or not conduct constitutes persecution in the Convention sense does not depend on the nature of the conduct. It depends on whether it discriminates against a person because of race, religion, nationality, political opinion or membership of a social group.[9]

---

[2] *Applicant A v Minister for Immigration and Ethnic Affairs* [1997] HCA 4, [1997] 190 CLR 225.
[3] *Horvath v Secretary of State for the Home Department* [2000] UKHL 37, [2001] AC 489 (Lord Hope).
[4] *Canada (Attorney General) v Ward* [1993] 2 SCR 689.
[5] Bundesverwaltungsgericht, *Beschluss vom 7. Februar 2008* – BVerwG 10 C 33.07, *S.J.L. v Refugee Appeals Tribunal and Others* [2014] IEHC 608, [43], *Matter of Maria Armida Sosa Ventura*, 25 I&N Dec. 391 (BIA 2010), United States Board of Immigration Appeals, 23 November 2010.
[6] *Matter of Acosta*, A-24159781, United States Board of Immigration Appeals, 1 March 1985.
[7] *Applicant A* (n 2) 248.
[8] Vienna Convention on the Law of Treaties (adopted 23 May 1969, entered into force 27 January 1980) 1155 UNTS 331.
[9] *Applicant A* (n 2) 258.

This broad characterisation incorporates denials of both civil and political as well as economic and social rights. It recognises the centrality of discrimination to the predicament faced by the claimant, a point that is further developed in Chapter 6. Although McHugh J does not address himself to the difference between 'persecution' and 'natural disasters', if his characterisation is taken as being to some extent indicative of what Dawson J had in mind when making his distinction between persecution and 'natural' disasters, it would suggest that 'natural' disasters lack the quality of discrimination that underpins all forms of persecution under the Convention. This construction does seem to reflect Dawson J's approach, as he explains later in his judgment:

> No doubt many of those limits in the present context spring from the well-accepted fact that international refugee law was meant to serve as a 'substitute' for national protection where the latter was not provided due to discrimination against persons on grounds of their civil and political status. It would therefore be wrong to depart from the demands of language and context by invoking the humanitarian objectives of the Convention without appreciating the limits which the Convention itself places on the achievement of them.[10]

Although there is some limited legal doctrinal work challenging this view,[11] the notion that 'natural disasters' do not discriminate is reflected at the highest levels of scholarship. Jane McAdam, for example, cites Dawson J's opinion in *Applicant A* after affirming that:

> even if the impacts of climate change could be characterized as 'persecution', the Refugee Convention requires such persecution to be *for reasons of* an individual's race, religion, nationality, political opinion, or membership of a particular social group. Persecution is not enough. The difficulty in the present context is that the impacts of climate change are largely indiscriminate, rather than tied to particular characteristics such as a person's background or beliefs.[12]

Interestingly, whereas Dawson J locates the requisite discriminatory quality of the refugee definition within the 'persecution' element, McAdam appears to see this quality as residing within the nexus to the five Convention reasons. Already, how the constitutive elements of the refugee definition are interpreted can be seen to have an impact on how RSD is conducted in this context. This point will be explored more closely in Chapter 6. For now, the epistemological point that people fleeing 'natural disasters' are not refugees because 'natural disasters' do not have the quality of discrimination that is integral to the notion of persecution for a Convention reason is in focus. This epistemology of disasters reflects the understanding articulated by Dr Jacob Robinson, Israeli Ambassador Plenipotentiary to the Refugee Convention at the Twenty-Second Meeting of the Conference of Plenipotentiaries:

> The text of sub-paragraph (2) obviously did not refer to refugees from natural disasters, for it was difficult to imagine that fires, floods, earthquakes or volcanic eruptions, for instance, differentiated between their victims on the grounds of race, religion or political opinion. Nor did that text cover all man-made events.[13]

---

[10] Ibid 248 (Dawson J). [11] See Section 1.3.
[12] Jane McAdam, *Climate Change, Forced Migration and International Law* (OUP 2012) 46.
[13] Conference of Plenipotentiaries on the Status of Refugees and Stateless Persons, Summary Record of the Twenty-Second Meeting (26 November 1951) UN Doc A/CONF.2/SR.22.

This view is reinforced in more contemporary times by António Guterres, then UN High Commissioner for Refugees:

> As the refugee definition only applies to those who have crossed an international border, the difficulties in characterising climate change as 'persecution', and the indiscriminate nature of its impacts, it does not expressly cover those fleeing a natural disaster or slow-onset degradation in living conditions owing to the environment.[14]

Thus, the dominant view in international refugee law is that the Refugee Convention rarely applies in the context of 'natural disasters' and climate change because these phenomena do not discriminate, and discrimination is an integral feature of the refugee definition. Closely related is the view that harm associated with 'natural disasters' cannot amount to 'persecution' because the necessary quality of human agency is missing. As McAdam frames this, '[p]art of the problem in the climate change context is identifying a "persecutor"'.[15]

This view that the discriminatory conduct of human actors of persecution is largely absent from the experience of disaster and climate-change-related harm appears to resonate with senior judges in other common law jurisdictions. Indeed, Dawson J's comment is quoted verbatim by Lord Hope in *Horvath*. Lord Hope is drawn to Dawson J's reasoning in another context where the limits of the Refugee Convention were required to be delineated. *Horvath* concerned the question of state protection and when a person could be said to have a well-founded fear of being persecuted in situations where the state was taking steps to protect the population from non-state actors. Recognising that the question the Court had to address went to the heart of the meaning of 'persecution', Lord Hope sought interpretative guidance in the underlying purpose of the Convention, which he construed thus:

> The general purpose of the Convention is to enable the person who no longer has the benefit of protection against persecution for a Convention reason in his own country to turn for protection to the international community.[16]

Understanding this purpose from the perspective articulated by Hathaway,[17] Lord Hope continued:

> If the principle of surrogacy is applied, the criterion must be whether the alleged lack of protection is such as to indicate that the home state is unwilling or unable to discharge its duty to establish and operate a system for the protection against persecution of its own nationals.[18]

It is in this connection that Lord Hope found assistance in the reasoning of Dawson J. Although Lord Hope is not concerned specifically with 'natural disasters', his adoption of Dawson J's reasoning in this context suggests a second distinction between 'natural disasters' and 'persecution'. In addition to the apparent lack of discrimination, people fleeing from 'natural' disasters would appear to still enjoy the protection of the state:

---

[14] UN High Commissioner for Refugees António Guterres, 'Migration, Displacement and Planned Relocation' (31 December 2012) fn 22 <http://www.unhcr.org/cgi-bin/texis/vtx/search?page=search&skip=342&docid=55535d6a9&query=water#_edn22>, accessed 4 April 2019.
[15] Jane McAdam, *Climate Change, Forced Migration and International Law* (OUP 2012), 45.
[16] *Horvath* (n 3) (Lord Hope).   [17] James Hathaway, *The Law of Refugee Status* (Butterworths 1991).
[18] *Horvath* (n 3), citing Hathaway, *Refugee Status* (n 17) 112.

I think that it follows that, in order to satisfy the fear test in a non-state agent case, the applicant for refugee status must show that the persecution which he fears consist [sic] of acts of violence or ill-treatment against which the state is unable or unwilling to provide protection. The applicant may have a well-founded fear of threats to his life due to famine or civil war or of isolated acts of violence or ill-treatment for a Convention reason which may be perpetrated against him. But the risk, however severe, and the fear, however well-founded, do not entitle him to the status of a refugee. The Convention has a more limited objective, the limits of which are identified by the list of Convention reasons and by the principle of surrogacy.[19]

What the approach of Lord Hope in *Horvath* reveals for the immediate purposes of this introductory chapter is the assumption that the state will 'do its best' in the context of 'natural disasters'; therefore, even when individuals risk dying in such a context, there is no need for surrogate international protection.[20]

Kälin and Schrepfer endorse this view in their policy paper prepared for UNHCR:

from a legal perspective, *there is an intrinsic difference between the two categories*: international refugee law is rooted in the notion of the surrogate nature of international protection ... [whereas] in the case of cross-border displacement caused by effects of climate change, the country of origin normally does not turn against affected people but remains willing to assist and protect them.[21]

This framing corresponds to Shacknove's formulation that the need for international protection arises when the 'bond of trust, loyalty, protection, and assistance between the citizen and the state' is severed.[22] Indeed, Shacknove summarises clearly the dominant view expressed by the judicial authorities highlighted above:

When determining who is, or is not, entitled to refugee status, natural disasters, such as floods and droughts, are usually dismissed as the bases for justified claims. Unlike the violent acts one person perpetrates against another, such disasters are not considered 'political' events. They are, supposedly, sources of vulnerability beyond social control which therefore impose no obligation on a government to secure a remedy. The bonds uniting citizen and state are said to endure even when the infrastructure or harvest of a region is obliterated. For even an ideally just state cannot save us from earthquakes, hurricanes, or eventual death.[23]

Clearly, the disaster imagined in these cases is intimately connected to the forces of nature. A drought does not discriminate and nor does a cyclone. State conduct is not implicated, nor is the conduct of non-state actors. Provided a state remains willing to assist victims in situations of disaster, such victims will not acquire refugee status upon crossing an international border. They fall outside of the refugee paradigm. With the combined legal

---

[19] Ibid 499 (Lord Hope).
[20] The approach to state protection adopted in *Horvath*, and reflected here, has been heavily criticised as being more concerned with process than result, and accepts that a person may have a well-founded fear of being persecuted but will nevertheless be unable to secure refugee status because the state has exercised due diligence. See James Hathaway and Michelle Foster, *The Law of Refugee Status* (2nd edn, CUP 2014) 314
[21] Walter Kälin and Nina Schrepfer, 'Protecting People Crossing Borders in the Context of Climate Change: Normative Gaps and Possible Approaches' (2012) UNHCR Legal and Protection Policy Research Series PPLA/2012/01, 31–32 (emphasis added) <http://www.unhcr.org/4f33f1729.pdf> accessed 4 April 2019.
[22] Andrew Shacknove, 'Who Is a Refugee?' (1985) 95 Ethics 274, 275.   [23] Ibid 279.

authority of senior courts, the UNHCR, and authoritative scholars of international refugee law, it is easy to see how the dominant view strongly supports the conclusion that the Refugee Convention has only peripheral relevance to the legal predicament of people displaced across borders in the context of 'natural' disasters and climate change.

1.3 A Different Perspective

The opinions cited above reflect a conception of people fearing exposure to climate- and disaster-related harm as being immediate victims of the storms, droughts, and floods whose indiscriminate impact causes adversity, which is the unfortunate consequence of the uncontrollable forces of nature, rather than the consequence of existing patterns of discrimination and marginalisation that generate unsafe conditions where individuals are exposed and vulnerable to natural hazard events. 'Natural disasters', according to this view, engender adversity. The role of human agents is entirely absent from the frame.

But Shacknove is alert to the operation of a different paradigm. Whereas one view is that disasters have nothing to do with violence and the state, an alternative view sees human agency pervading 'natural' disasters:

> But as writers such as Lofiche, Sen, and Shue have demonstrated, 'natural disasters' are frequently complicated by human actions. The devastation of a flood or a supposedly natural famine can be minimized or exacerbated by social policies and institutions. As Lofchie says: 'The point of departure for a political understanding of African hunger is so obvious it is almost always overlooked: the distinction between drought and famine ... To the extent that there is a connection between drought and famine, it is mediated by the political and economic arrangements of society. These can either minimize the human consequences of drought or accentuate its effects'.[24]

From the above it would appear that to answer Shacknove's question of 'who is a refugee?' in the context of 'natural' disasters and climate change, it is necessary not only to understand the meaning of the inclusion criteria in Article 1A(2), but also to have a clear conception of what a 'natural' disaster actually is. Where disasters are equated with natural hazards, people displaced in that connection may appear to fall outside of the refugee law paradigm. However, when disasters are understood from a more social perspective, the relevance of the Refugee Convention to the predicament of displaced persons appears far less remote.

Other legal doctrinal scholars have examined the significance of a clear focus on the social context in which disasters and environmental degradation take place. Although his engagement is not as far reaching as others, Hathaway's treatment of the application of the Refugee Convention in the context of disasters warrants attention because it is amongst the earliest. It has also been incompletely construed by some judicial authorities, who cite *The Law of Refugee Status* as academic authority for the proposition that a person fleeing in the context of a 'natural disaster' cannot establish eligibility for recognition of refugee status. In the section 'assessing risk within the context of generalised oppression', Hathaway approves of the view expressed by Robinson above, explaining:

---

[24] Ibid

> The rationale for considering the victims of natural disasters or widespread turmoil to be outside the scope of the Convention is not, however, their adverse impact on large numbers of persons, but is rather the non-discriminatory nature of the risk.[25]

However, although some of the judicial authorities that will be surveyed in Chapter 3 stop at this point, Hathaway's approach is not as unequivocal as the above statement would suggest. He instead explains that, in accordance with his general human-rights-based approach to RSD, a person should establish eligibility for refugee status when, owing to a discriminatory state response to a situation of disaster, a person is exposed to serious harm:

> By way of example, the victims of a flood or earthquake are not *per se* Convention refugees, even if they have fled to a neighbouring state because their own government was unable or unwilling to provide them with relief assistance. If, on the other hand, the government of the home state chose to limit its relief efforts to those victims who were members of the majority race, forcing a minority group to flee to another country in order to avoid starvation or exposure, a claim to refugee status should succeed because the harm feared is serious and connected to the state, and the requisite element of civil and political differentiation is present.[26]

Hathaway develops his approach further in a much later article focusing on food deprivation, developing the argument that:

> Because the link to a 'failure' of state protection is all that must be shown under the bifurcated approach, it is sufficient to show that the government simply could not be bothered to protect a portion of the at-risk group – reasoning, for example, that because they are 'only' women or indigenous persons they were not worthy of an expenditure of government resources. In such circumstances, the failure of protection is causally connected to a Convention ground and refugee status should be recognised.[27]

Failures of state protection even before a disaster unfolds may also have a bearing on a claim for recognition of refugee status. Kolmannskog identifies that marginalised groups may be differentially exposed and vulnerable to disaster-related harm, and that this fact may have significance for RSD:

> Certain marginalised groups of people such as ethnic minorities and political dissidents are often more vulnerable and exposed to disasters in the first place and receive less protection and assistance during and after a disaster. In a similar manner to gender cases, the nexus requirement of the Convention could be fulfilled when lack of protection from the state is linked to one of the five grounds (race, religion, nationality, membership of a particular social group or political opinion).[28]

Burson also takes pains to emphasise this social and historical context in which displacement in the context of environmental degradation and disasters needs to be understood in order for RSD to be safely performed. Expressly identifying the danger of judicial comments

---

[25] Hathaway, *Refugee Status* (n 17) 92–93.   [26] Ibid 94.
[27] James Hathaway, 'Food Deprivation: A Basis for Refugee Status?' (2014) 81 Soc Res 327, 336 <http://repository.law.umich.edu/articles/1076/> accessed 4 April 2019.
[28] Vikram Kolmannskog, 'Climate Change, Environmental Displacement and International Law' (2012) 24 J Int Dev 1071, 1076.

'that persons "fleeing natural disasters" cannot obtain Convention-based protection ... being relied on more broadly than warranted',[29] he explains:

> Environmental degradation is intimately bound up with long-term issues of development, population growth, and economic and social policy choices ... This is particularly true in relation to climate change ... This historical context, when mixed with activity of a discriminatory nature, can in principle produce environmentally displaced persons who meet the Convention's definition.[30]

Unfortunately, the full implications of this clear adoption of what in Chapter 2 will be described as the 'social paradigm' are not explored, as the scenarios Burson identifies as potentially engendering eligibility for refugee status are familiar examples of violence enacted against a backdrop of disasters and climate change.[31]

At the same time, Burson argues that RSD ought not focus exclusively on *violence*, as 'contemporary approaches to this issue may bring more environmentally displaced persons within the scope of the Convention than those where violence is imminent'.[32] Following Hathaway, Burson argues that:

> [a]t the heart of the Convention is the avoidance of discrimination in the enjoyment of basic human rights. Where the discrimination encountered in the context of environmental degradation interferes with the basic human rights of a marginalised group, this provides a clear avenue by which the circumstances of the marginalised group may fall within the scope of the Convention. This is not to say that they automatically will. The legal condition of 'being persecuted' connotes a particular form of discrimination. Whether their circumstances reach the Convention's threshold will depend on the nature and extent of the discrimination encountered.[33]

Clearly countering the view that environmental degradation or disasters have an indiscriminate impact, Burson expressly invites decision-makers to examine the social context, including in situations of mass-influx, recognising that certain individuals amongst a wider group may satisfy the eligibility requirements for recognition of refugee status.[34]

1.4 Structure of the Book

Having in this introductory chapter identified how important it is to understand what disasters actually are, Chapter 2 presents the two competing 'hazard' and 'social' paradigms. Chapter 3 then presents the findings of a review of jurisprudence addressing actual claims for recognition of refugee status in the context of disasters and climate change. This chapter highlights how both epistemological and doctrinal assumptions affect how refugee status is determined on a case-by-case basis. Finding much to commend in the clear and principled

---

[29] Bruce Burson, 'Environmentally Induced Displacement and the 1951 Refugee Convention: Pathways to Recognition', in T Afifi and J Jäger (eds), *Environment, Forced Migration and Social Vulnerability* (Springer 2010) 9.
[30] Ibid 7.
[31] Burson notes, for example, that states intentionally destroy certain environments in order to cause harm to certain groups, as occurred under Saddam Hussein's campaign against the Marsh Arabs. He further notes scenarios where activists protesting against failures by the state to prevent environmental degradation are singled out for ill treatment by the regime. See ibid 7.
[32] Ibid 8. [33] Ibid 10. [34] Ibid 9.

human-rights-based approach adopted by the New Zealand Immigration and Protection Tribunal, Chapters 4–6 are devoted to what appear as limitations in the approach, from whence a recalibrated human-rights-based interpretation of the refugee definition is proposed. Finally, Chapter 7 examines how this recalibrated interpretation, informed by the social paradigm developed in Chapter 2, may be applied to claims for recognition of refugee status in the context of disasters and climate change.

2

Two Disaster Paradigms

2.1 What Is a Disaster?

The definition of 'climate change' is provided at Article 1(2) of the UN Framework Convention on Climate Change:

> 'Climate change' means a change of climate which is attributed directly or indirectly to human activity that alters the composition of the global atmosphere and which is in addition to natural climate variability observed over comparable time periods.[1]

The process of climate change takes place one step removed from the impacts of climate change, which can manifest, inter alia, in increasingly frequent and intense natural hazard events. Hence, this phenomenon is better understood as a threat or impact multiplier,[2] rather than being a tangible risk that a person can be exposed and vulnerable to.

This book therefore focuses primarily on the phenomenon of the 'natural' disaster whilst remaining acutely aware of the impact multiplier role played by ongoing and anticipated future changes to the climate. This brief subsection will not canvass the myriad definitions, nor attempt a typology, examples of which can be found elsewhere.[3] Rather, the definition adopted by the International Law Commission in the Draft Articles on the Protection of Persons in Situations of Disaster is selected as an important legal definition that arose in the course of a wide consultative process. The definition is interesting because it also reflects key features of the hazard paradigm, and many states commenting on the Draft Articles have highlighted its limitations with reference to insights from the social paradigm.

The International Law Commission's Draft Articles on the Protection of Persons in Situations of Disaster defines a disaster as:

---

[1] UN Framework Convention on Climate Change (adopted 9 May 1992, entered into force 21 March 1994) 1771 UNTS 107.

[2] See for example Christopher Field et al (eds), *Managing the Risks of Extreme Events and Disasters to Advance Climate Change Adaptation: Special Report of the Intergovernmental Panel on Climate Change* (CUP 2012) 458; Jane McAdam, *Climate Change, Forced Migration and International Law* (OUP 2012) 24; Norwegian Refugee Council/Internal Displacement Monitoring Centre (NRC/IDMC), 'The Nansen Conference: Climate Change and Displacement in the 21st Century' (7 June 2011) 18 <http://www.refworld.org/docid/521485ef4.html> accessed 4 April 2019.

[3] See for example David Alexander, 'The Study of Natural Disasters, 1977–1997: Some Reflections on a Changing Field of Knowledge' (1997) 21 Disasters 284; Ronald Perry, 'What Is a Disaster?', in Havidán Rodríguez, Enrico Quarantelli and Russell Dynes (eds), *Handbook of Disaster Research* (Springer 2007).

a calamitous event or series of events resulting in widespread loss of life, great human suffering and distress, mass displacement, or large-scale material or environmental damage, thereby seriously disrupting the functioning of society.[4]

That this definition determines the scope of such a significant document as the Draft Articles highlights the continued epistemological resonance of the hazard paradigm. The emphasis here is on (a) large-scale (b) events (c) that have serious human or material impacts that (d) seriously disrupt the functioning of society. The Commentary to the Draft Articles confirms the intention to delimit the definition to events, rather than more complex processes, and to extremes rather than less dramatic events, so as to ensure the precision of the definition, even at the expense of what it considers more 'policy-oriented' insights.[5] The decision not to identify the source of the event was taken to enable disasters to refer to events arising both from natural hazards as well as anthropogenic triggers.

This definition received criticism from some states, inter alia, for its failure to reflect contemporary understandings of disasters. The United States, for example, stated that it had:

> significant concerns with the Commission's proposed definition of 'disaster' in Draft Article 3. First, we question the decision to define disaster in terms of an 'event,' rather than in terms of the consequences of an event combined with vulnerable social conditions. As the Commentary notes, the majority of the nonbinding instruments that specifically address disasters focus on the types of hazards and social conditions of vulnerability that disrupt the normal functioning of a community or society. Further, since the first reading of these Draft Articles, States have adopted the nonbinding Sendai Framework for Disaster Risk Reduction, which also focuses on hazards, vulnerability, and risks, and the Commission should take into consideration that broadly negotiated framework.[6]

Similarly, the intentional delimitation of the definition to 'events', as distinct from 'processes', was challenged by Germany:

> The definition of 'disaster' should not only focus on fast-onset 'events', but also on slow-onset processes such as droughts, which pose a huge threat to high-risk countries. We therefore propose that 'prolonged processes' be incorporated into the definition of a disaster in draft article 3.[7]

The decision to focus on events 'seriously disrupting the functioning of society' received criticism from Austria, which observed that the impact of earthquakes, tsunamis, floods, and other hazard events would not necessarily attain that threshold:

> It is doubtful whether an earthquake, an avalanche, a flood or a tsunami taken as such necessarily meets the threshold of a 'serious disruption of society'. If the present definition were taken literally, situations as frequent as those – and expected to fall within the envisaged ambit – would not always be classified as disasters for the purposes of the draft articles.[8]

---

[4] ILC, *Draft Articles on the Protection of Persons in the Event of Disaster, Report of the Work of the 68th Session* (2 May–10 June and 4 July–12 August 2016) UN Doc A/71/10, Article 3(a), 14.
[5] Ibid 22–24.
[6] ILC, *Protection of Persons in the Event of Disasters: Comments and Observations Received from Governments and International Organizations* (28 April 2016) UN Doc A/CN.4/696/Add.1, 6.
[7] Ibid [8] Ibid 13.

A similar view was expressed by the Secretariat of the UN Office for Disaster Risk Reduction, which made particular reference to 'small-scale disasters'.[9]

From these comments it is clear that the definition set out in the Draft Articles does not reflect contemporary understandings of what 'natural' disasters are. Consequently, although the Commentary[10] reflects an awareness of more state of the art approaches, the definition of 'disaster' at Article 3(a) risks reproducing the 'hazard' paradigm that was identified as operating in the thinking of judicial decision-makers and scholars of international refugee law surveyed in the introduction. The hazard paradigm retains a powerful grip, despite no longer reflecting contemporary understanding within the disaster risk reduction community.

2.2 The Hazard Paradigm

It is not surprising that hazards such as cyclones, floods, and droughts become, in the term 'natural disaster', synonymous with the disasters they can trigger. Over the course of just a few weeks in 2017, the US state of Texas had been flooded by Hurricane Harvey,[11] monsoonal rains had flooded large swathes of South Asia,[12] and Caribbean islands had been devastated by Hurricane Irma, the strongest hurricane ever recorded in the Atlantic.[13] Commentators reflect on how climate change may have contributed to these events,[14] and anticipate worse to come.[15] To be clear, the hazards that bring downpours, powerful winds, and storm surges present serious threats that cannot be totally eliminated. Indeed, the intensifying impact of climate change presents an existential challenge to humanity.

However, the uncritical use of the term 'natural disaster' reflects the continued influence of an approach to the understanding of 'natural' disasters that focuses on the hazard independently of the social context. As Perry notes, 'although there may be a concern with social and other issues, the real emphasis is on the processes associated with the target agent'.[16]

The hazard paradigm is characterised by Hewitt in his 1983 classic as 'the dominant view':

> The superficial features of the dominant view are not hard to discern ... There is generally a straightforward acceptance of natural disasters as a result of 'extremes' in

---

[9] Ibid 14. [10] ILC, *Report of the Work of the 68th Session* (n 4).
[11] Niraj Chokshi and Maggie Astor, 'Hurricane Harvey: The Devastation and What Comes Next' *New York Times* (28 August 2017) <https://www.nytimes.com/2017/08/28/us/hurricane-harvey-texas.html> accessed 4 April 2019.
[12] Simon Mundy, 'Heavy Flooding Kills More than 1,000 in South Asia' *Financial Times* (30 August 2017) <https://www.ft.com/video/afc1b141-ad96-4084-875c-185d69e62a7c> accessed 4 April 2019.
[13] Gregor Aisch, Adam Pearce and Karen Yourish, 'Hurricane Irma Is One of the Strongest Storms in History' *New York Times* (9 September 2017) <https://www.nytimes.com/interactive/2017/09/09/us/hurricane-irma-records.html> accessed 4 April 2019.
[14] See for example Bill McKibben, 'Stop Talking Right Now about the Threat of Climate Change: It's Here; It's Happening' *The Guardian* (11 September 2017) <https://www.theguardian.com/commentisfree/2017/sep/11/threat-climate-change-hurricane-harvey-irma-droughts> accessed 4 April 2019.
[15] Jonathan Watts, 'Twin Megastorms Have Scientists Fearing This May Be the New Normal' *The Guardian* (6 September 2017) <https://www.theguardian.com/world/2017/sep/06/twin-megastorms-irma-harvey-scientists-fear-new-normal> accessed 4 April 2019.
[16] Perry, 'What Is a Disaster?' (n 3) 8–9.

geophysical processes. The occurrence and essential features of calamity are seen to depend primarily upon the nature of storms, earthquakes, flood, drought.[17]

This 'dominant view' does not, however, ignore the fact that social factors play a role in 'natural disasters'. The difference lies in the primacy attached to the hazard, which is seen to determine how social conditions may be relevant:

> Few researchers would deny that social and economic factors or habitat conditions other than geophysical extremes affect risk. The direction of argument in the dominant view relegates them to a dependent position. The initiative in calamity is seen to be with nature, which decides where and what social conditions or response will become significant.[18]

Although nature has the initiative in this paradigm, states can protect populations through technocratic intervention:

> In the dominant view, then, disaster itself is attributed to nature. There is, however, an equally strong conviction that something can be done about disaster by society. But that something is viewed as strictly a matter of public policy backed up by the most advanced geophysical, geotechnical and managerial capability.[19]

The fact that a state may fail to protect a population from such forces of nature is clearly unfortunate, but there is little in this paradigm that would suggest that governance, or other social factors, might play a causal, as opposed to a reactive, role in the unfolding of 'natural disasters'.

Cedervall Lauta devotes the opening of his study on disaster law to a discussion of how disasters have been conceptualised within the European tradition.[20] He identifies the Lisbon earthquake of 1755 as the turning point at which the then dominant paradigm that saw disasters as acts of God came instead to be understood as the consequence of natural hazards that were capable of being studied and, to some extent, mitigated against.[21] He writes:

> the sciences of seismology, meteorology and volcanology have developed with the purpose of describing and investigating the contingent nature and thereby mitigating the damage if or when a natural hazard was edging closer. The main focus of disaster management was not prevention, but mitigation and the increasing understanding of these hazards.[22]

Hilhorst describes the limited attention afforded under this 'hazard' model to the social context in which disasters unfold:

> It coupled a hazard-centred interest in the geo-physical processes underlying disaster with the conviction that people had to be taught to anticipate disaster. It is a technocratic paradigm dominated by geologists, seismologists, meteorologists and other scientists who can monitor and predict the hazards, while social scientists are brought in to

---

[17] Kenneth Hewitt, 'The Idea of Calamity in a Technocratic Age' in K Hewitt (ed), *Interpretations of Calamity from the Viewpoint of Human Ecology* (Allen and Unwin 1983) 5.
[18] Ibid [19] Ibid 6. [20] Kristian Cedervall Lauta, *Disaster Law* (Routledge 2015) ch 2.
[21] Ibid 12–13. [22] Ibid 19.

explain people's behaviour in response to risk and disaster and develop early warning mechanisms and disaster preparedness schemes.[23]

Hewitt also describes the 'hazard paradigm' as being outdated and inaccurate, despite retaining a powerful hold on the imagination and in practice:

> The danger lies with, in Gilbert's terms, a 'hazards paradigm' – a viewpoint that classifies, explains and responds to disasters as if they were wholly or essentially a function of the agent that 'impinges upon a vulnerable society'. This paradigm renders social understanding secondary if not impossible, by placing the sources of risk literally outside society, 'in the environment,' or presumed accidental, unscheduled forces that 'erupt' within ... Society – at least, communities, the public or populations – are made to appear passive victims of natural and technological agents.[24]

Thus, the hazard paradigm elides the notion of the disaster with the occurrence of the triggering hazard event. The social paradigm that is adopted below does not in any way deny the destructive forces of nature. Rather, it invites consideration of how these forces interact with exposed and vulnerable social conditions in the unfolding of disaster.

Bankoff explains the narrow temporal scope of the disaster when understood from within the hazard paradigm:

> The traditional explanation of a disaster as an abnormal geophysical or meteorological event confers a certain cyclical progression upon its incidence as societies move through a series of stages from quiescence, to warning, pre-alarm, crisis, rescue, rehabilitation and finally reconstruction ... However, the disaster itself always has a limited temporal existence in that it has an identifiable commencement and a recognised cessation. The assumptions are that it is unexpected, its onslaught sudden, its duration brief, its effects uniform and its repercussions localised ... Disasters are regarded as the work of god or nature, that is they are 'natural disasters', and the frequency of their occurrence largely a matter of chance.[25]

A hazard paradigm that imputes agency to the forces of nature and that zeroes in on the occurrence of the hazard event itself as if it were indistinguishable from the disaster that unfolds in that connection is clearly unhelpful for the purpose of refugee status determination (RSD). A social paradigm is superior for explaining how hazard events become disasters, and why some people may be vulnerable and exposed to disaster-related harm for reasons of race, religion, nationality, membership of a particular social group, or political opinion.

That the hazard paradigm as described above fails to adequately describe *disasters* because of its inadequate engagement with the social factors that engender exposure and vulnerability to natural hazards does nothing to diminish the importance of the cyclone, earthquake, or flood. 'Natural' disasters cannot happen without a natural hazard event or

---

[23] Dorothea Hilhorst, 'Unlocking Disaster Paradigms: An Actor-Oriented Focus on Disaster Response' (Abstract Submitted for Session 3 of the Disaster Research and Social Crisis Network Panels of the 6th European Sociological Conference, 23–26 September, Murcia, Spain, 2003) 2; available on request from <https://www.researchgate.net/publication/254033796_Unlocking_disaster_paradigms_An_actor-oriented_focus_on_disaster_response> accessed 17 April 2019, citing Anthony Oliver-Smith, 'Anthropological Research on Hazards and Disasters' (1996) 25 Annu Rev Anthropol 303.

[24] Kenneth Hewitt, 'Excluded Perspectives in the Social Construction of Disaster' (1995) 13 IJMED 317, 320.

[25] Greg Bankoff, *Cultures of Disaster: Society and Natural Hazard in the Philippines* (RoutledgeCurzon 2003) 155–56.

process unfolding, just as a disaster will not unfold where people are not exposed and vulnerable. In embracing the social paradigm described below, this book seeks to draw out the social context in which 'natural' disasters unfold, because this frame is of most direct relevance to refugee status determination.

2.3 The Social Paradigm

The social paradigm differs from the hazard paradigm in three ways that are relevant to identifying the kinds of circumstances in which a person may establish a well-founded fear of being persecuted for a Convention reason in the context of 'natural' disasters and climate change. First, the social paradigm sees 'natural' disasters as a consequence of the interaction of natural hazards and social vulnerability. Consequently, human agency is inherent in all 'natural' disasters. Second, it recognises that within this social context, certain individuals may be more vulnerable than others on account of pre-existing patterns of discrimination. Hence, 'natural' disasters do not have an indiscriminate impact. Third, the social paradigm understands 'natural' disasters as process, in the sense that individual and societal vulnerability and exposure to natural hazard events is historically contingent and changes over time.

The widespread notion of the 'natural disaster' therefore falls to be disaggregated. The occurrence of a natural hazard event or process is a necessary, but not sufficient, condition for the unfolding of a 'natural' disaster. Without exposed and vulnerable human settlements, a flood will not engender disaster.

Whilst not seeking to replace the natural-scientific study of floods, earthquakes, cyclones, and so forth, social scientists working under themes such as disaster anthropology and political ecology have, through their identification and analysis of historical, cultural, social, political, economic, and other factors, transformed the study of 'natural' disasters by 'taking the naturalness out of natural disasters'. A call issued as long ago as 1976, under a journal article by the same name, argued:

> The time is ripe for some form of precautionary planning which considers vulnerability of the population as the real cause of disaster – a vulnerability that is induced by socio-economic conditions that can be modified by man, and is not just an act of god. Precautionary planning must commence, with the removal of concepts of naturalness from natural disasters.[26]

Describing his anthropological approach, which draws on political ecology, Oliver-Smith explains:

> Implicit in my approach is the assumption that disasters are as deeply embedded in the social structure and culture of a society as they are in an environment. In a sense, a disaster is symptomatic of the condition of a society's total adaptational strategy within its social, economic, modified, and built environments.[27]

---

[26] Phil O'Keefe, Ken Westgate and Ben Wisner, 'Taking the Naturalness Out of Natural Disasters' (1976) 260 Nature 566, 567.

[27] Oliver-Smith, '"What Is a Disaster?" Anthropological Perspectives on a Persistent Question' in A Oliver-Smith and S Hoffman (eds), *The Angry Earth: Disaster in Anthropological Perspective* (Psychology Press 1999) 25.

He adds:

> A political ecology perspective on disasters focuses on the dynamic relationships between a human population, its socially generated and politically enforced productive and allocative patterns, and its physical environment, all in the formation of patterns of vulnerability and response to disaster.[28]

Appraising the impact of what he describes as this 'social turn' in disaster studies,[29] Cedervall Lauta writes:

> Rather than understanding the disaster as an external, intruding force into society, it could be understood as a phenomenon (partially) produced by the society itself. The disaster is in this paradigm not only understood as an intervening disturbance, but a phenomenon generated from our own system's weaknesses. Thus, the paradigm changes what we would previously [have] conceived of as bad luck into deficiencies and repudiates nature from the centre of disaster studies to a characterising component of a fundamentally social phenomenon. The paradigm thereby radically transforms the epistemology of disaster.[30]

Many specialists in the field of disaster risk reduction would agree with Cedervall Lauta's characterisation of this new epistemology of disaster. Sálvano Briceño, former Director of the UN Office for Disaster Risk Reduction, laments:

> The confused use of the phrase 'natural disaster' is still common. This misperception, currently encouraged by the climate change spotlight, has led to an excessive focus on understanding the hazards themselves from a physical and natural sciences point of view, and does not facilitate the much-needed attention to understanding and reducing human and social vulnerability.[31]

The social paradigm, then, provides for a more comprehensive understanding of the constitutive elements of a 'natural' disaster than the hazard paradigm. Central to the social paradigm is the notion of vulnerability.

2.3.1 Vulnerability

According to the social paradigm, disasters happen as a consequence of the interaction of hazards with vulnerable social conditions. The concept is often communicated as a pseudo-equation: $DR = H \times V$ (disaster risk = hazard × vulnerability).[32]

---

[28] Ibid 29–30.   [29] Cedervall Lauta, *Disaster Law* (n 20) 20.   [30] Ibid
[31] Sálvano Briceño, 'Forward' in Ben Wisner, JC Gaillard and Ilan Kelman (eds), *The Routledge Handbook of Hazards and Disaster Risk Reduction* (Routledge 2012) xxxii.
[32] The view that disaster results from a combination of natural hazards, vulnerability, and exposure is prominent in the literature on disaster risk reduction. See for example Field et al (eds), *Managing the Risks of Extreme Events and Disasters* (n 36); Ben Wisner et al, *At Risk: Natural Hazards, People's Vulnerability and Disasters* (2nd edn, Routledge 2004); UNISDR/ESCAP, *Reducing Vulnerability and Exposure to Disasters: The Asia Pacific Disaster Report 2012* (UNISDR 2012) xxi; IFRC, *What Is a Disaster?* <https://www.ifrc.org/en/what-we-do/disaster-management/about-disasters/what-is-a-disaster/> accessed 4 April 2019; Cedervall Lauta, *Disaster Law* (n 20).

That this perspective has become mainstream is nowhere more in evidence than at paragraph 6 of the flagship Sendai Framework on Disaster Risk Reduction, which emphasises:

> 6. Enhanced work to reduce exposure and vulnerability ... More dedicated action needs to be focused on tackling underlying disaster risk drivers, such as the consequences of poverty and inequality, climate change and variability, unplanned and rapid urbanization, poor land management and compounding factors.[33]

As hazards themselves cannot be prevented, key aims of disaster risk reduction include reducing exposure and vulnerability and increasing resilience.[34]

Exposure refers to the physical proximity to hazard. Houses built on steep mountain slopes are exposed to earthquakes and landslides. Slum settlements built along river banks are exposed to flooding. Coastal settlements are exposed to sea level rise, storm surges, tropical storms, and so forth.

Exposure to natural hazards is not synonymous with vulnerability, although the two are closely correlated. The UN Office for Disaster Risk Reduction (since 2019 referred to as UNDRR) updated its definition of vulnerability in 2017 to read: 'the conditions determined by physical, social, economic and environmental factors or processes which increase the susceptibility of an individual, a community, assets or systems to the impacts of hazards'.[35] This change represents a significant deepening of the appreciation of how disasters can have differential impacts even within the same community. The earlier definition described vulnerability as 'the characteristics and circumstances of a community, system or asset that make it susceptible to the damaging effects of a hazard'.[36]

At the community level, vulnerability to natural hazard events might entail:

> poor design and construction of buildings, inadequate protection of assets, lack of public information and awareness, limited official recognition of risks and preparedness measures, and disregard for wise environmental management.

Clearly, vulnerability can only be understood in its social context.

However, by focusing on 'community, system or asset', what the earlier definition and explanation of vulnerability failed to provide is insight into how vulnerability is differentially experienced across a population. Some buildings may be poorly constructed, whereas others may be more robust. Some people may have access to information, whilst others may not. Likewise, other characteristics of vulnerability – such as precarious livelihood activities, individual levels of health, wealth, education, social capital, and so forth – are not evoked in this definition.

The refugee law practitioner needs to go beyond perceptions of vulnerability that hover above the surface of social relations if questions relevant to RSD are to be asked and answered in a sufficiently robust way. Thus, for the refugee lawyer, instead of thinking about 'the community' in a setting that is exposed to a particular hazard, questions need to be asked about particular groups and their distinctive characteristics, their relations with other groups, and how they have come to occupy particular niches in different localities.

---

[33] Sendai Framework on Disaster Risk Reduction 2015–2030, A/CONF.224/CRP.1.   [34] Ibid
[35] UNDRR, *Terminology on Disaster Risk Reduction* <http://www.unisdr.org/we/inform/terminology> accessed 12 August 2019.
[36] UNISDR, *Terminology on Disaster Risk Reduction* (May 2009) <http://www.unisdr.org/files/7817_UNISDRTerminologyEnglish.pdf> accessed 4 April 2019.

This disaggregated approach to notions of 'community' is endorsed by Peacock and Ragsdale, who reject 'the notion of community as a single autonomous social system', arguing:

> a community is an ecological network of groups and organizations linked through divisions of labor based on contingent relationships. Competition and conflict are inherent ... Finally, differential access to network resources is critical for understanding the survival and reproduction of its social units.[37]

Similarly, Guijt and Shah argue:

> Inequalities, oppressive social hierarchies and discrimination are often overlooked, and instead enthusiasm is generated for the cooperative and harmonious ideal promised by the imagery of 'community'.[38]

Drawing on his experience in India, Cannon reveals clearly how 'communities' are places of 'unequally distributed risks and vulnerabilities':

> In the place we want to call a community, some have better opportunities than others, some are richer than others, and some are safer than others. In the flood-prone north Indian villages I have visited, the outcaste untouchables (now often referred to as Dalits) live in squalid, flood-prone settlements on the low-lying edges of villages, while the better-off have homes that are on raised land at the centre ... So to understand vulnerability is to understand community not as a harmonious place with the potential for mutuality and risk sharing, but a place of unequally distributed risks and vulnerabilities.[39]

UNDRR's new focus on the individual in social context is in clear contrast to the community-based focus reflected in the older definition.

Having now argued in general terms that certain individuals and groups may be more vulnerable and exposed to disaster-related harm than others, it is necessary to consider what factors underpin such differential exposure and vulnerability.

2.3.2 The Pressure and Release Model

The 'Pressure and Release' (PAR) model is 'the most frequently cited theoretical model'[40] for understanding 'the progression of vulnerability', and it is expressly based on the recognition of the differential impact of disasters. Its theoretical framework informs contributions by over 70 leading scholars and practitioners to the 2012 *Routledge Handbook of*

---

[37] Walter Gillis Peacock with A. Kathleen Ragsdale, 'Social Systems, Ecological Networks and Disasters: Toward a Socio-Political Ecology of Disasters' in Walter Gillis Peacock, Betty Hearn Morrow and Hugh Gladwin (eds), *Hurricane Andrew: Ethnicity, Gender and the Sociology of Disasters* (Routledge 1997) 24.

[38] Irene Guijt and Meera Kaul Shah, *The Myth of Community: Gender Issues in Participatory Development* (Intermediate Technology Publication 1998) 7–8.

[39] Terry Cannon, 'Reducing People's Vulnerability to Hazards: Communities and Resilience' (2008) United Nations University – WIDER Research Paper No. 2008/34, 13 <https://www.wider.unu.edu/publication/reducing-people's-vulnerability-natural-hazards> accessed 4 April 2019.

[40] Susan L. Cutter et al, *Social Vulnerability to Climate Variability Hazards: A Review of the Literature* (2009) Final Report to Oxfam America, 5 <http://citeseerx.ist.psu.edu/viewdoc/download?doi=10.1.1.458.7614&rep=rep1&type=pdf> accessed 4 April 2019.

Hazards and Disaster Risk Reduction.[41] The approach, developed by Wisner et al in their classic *At Risk: Natural Hazards, People's Vulnerability and Disasters*,[42] describes vulnerability as the culmination of root causes, dynamic pressures, and unsafe conditions. Under this PAR model, vulnerability is understood as:

> the degree to which one's social status (e.g. culturally and socially constructed in terms of roles, responsibilities, rights, duties and expectations concerning behaviour) influences differential impact by natural hazards and the social processes which led there and maintain their status.[43]

Understanding the *root causes* of vulnerability can entail the exploration of centuries of social history,[44] but in more practical terms it invites an appreciation of the structural underpinnings of a society. People do not simply end up living in places that are exposed to natural and other hazards, and they do not simply happen to lack the resilience to protect themselves from such hazards and recover in the aftermath. There is a story to be uncovered. Wisner et al explain:

> The most important root causes that give rise to vulnerability (and which reproduce vulnerability over time) are economic, demographic and political processes. These affect the allocation and distribution of resources, among different groups of people. They are a function of economic, social, and political structures, and also legal definitions and enforcement of rights, gender relations and other elements of the ideological order.[45]

Dynamic pressures are those contemporary factors that affect how people live in a society. They include such factors as government policies, for example relating to land distribution, education, and international trade, as well as levels of corruption, the presence or absence of social unrest and/or armed conflict, and so forth. Dynamic pressures

> are more contemporary or immediate, conjunctural manifestations of general underlying economic, social and political patterns ... Dynamic pressures channel the root causes into particular forms of unsafe conditions that then have to be considered in relation to the different types of hazards facing people. These dynamic pressures include epidemic disease, rapid urbanisation, current (as opposed to past) wars and other violent conflicts, foreign debt and certain structural adjustment programmes.[46]

An appreciation of root causes and dynamic pressures underpinning unsafe conditions invites an approach to disasters as *processes*, as opposed to isolated, exceptional *events*. Expressing a perspective on disasters that converges with the PAR model, Oliver-Smith explains:

---

[41] Ben Wisner, JC Gaillard and Ilan Kelman (eds), *The Routledge Handbook of Hazards and Disaster Risk Reduction* (Routledge 2012).

[42] Wisner et al, *At Risk* (n 32) ch 2.

[43] Ben Wisner, JC Gaillard and Ilan Kelman, 'Framing Disaster' in Wisner et al (eds), *Handbook of Hazards* (n 41) 22.

[44] Consider Michael Watts, *Silent Violence: Food, Famine and Peasantry in Northern Nigeria* (University of California Press 1983), which traces contemporary vulnerability to drought and related famine in northern Nigeria back over 200 years, and Anthony Oliver-Smith, 'Haiti's 500-Year Earthquake' in Mark Schuller and Pablo Morales (eds), *Tectonic Shifts: Haiti since the Earthquake* (Kumarian Press 2012).

[45] Wisner et al, *At Risk* (n 32) 52.   [46] Ibid 54.

The basic view is that a necessary but not sufficient condition for a disaster to occur is the conjuncture of at least two factors: a human population and a potentially destructive agent. The society and the destructive agent are mutually constitutive and embedded in natural and social systems as unfolding processes over time. Both societies and destructive agents are clearly processual phenomena, together defining disaster as a processual phenomenon rather than an event that is isolated and temporally demarcated in exact time frames.[47]

Similarly, Bankoff articulates how 'vulnerability may take centuries in the making', by recalling Peru's '500-year earthquake' in the 1970s and the 'structural role played by external and internal colonialism as factors in determining those disasters'. He further highlights how '[c]ertain segments of a population are often situated in more perilous settings than others due to the historical consequences of political, economic and/or social forces'.[48]

Recognising disasters as processes discourages facile recourse to distinctions between 'sudden-onset' and 'slow-onset' disasters, as Twigg explains:

To some extent, the distinction between slow- and rapid-onset disasters is artificial. Hazards certainly can be categorised in this way. Disasters, on the other hand, are the product of hazards and human vulnerability to them. The socio-economic forces that make people vulnerable may act quickly or slowly, but in most disasters it is likely that long-term trends will be more influential. When viewed in this light, it could be argued that all disasters are slow-onset.[49]

This understanding of 'natural' disasters as processes has relevance for questions around whether a person's fear of being persecuted for a Convention reason is well founded, and will be discussed further in Chapter 5.

The *unsafe conditions* aspect of the PAR model concerns the individual or community position, understood within the context of those root causes and dynamic pressures. Why do certain people live in a particular place, and what are the factors that conspire to make that place one that is exposed to natural hazards? Additionally, what makes this individual or community vulnerable, in terms of protecting against hazards and rebuilding their lives and livelihoods in the aftermath?

Summing up the model, Wisner et al write:

most people are vulnerable because they have inadequate livelihoods, which are not resilient in the face of shocks, and they are often poor. They are poor because they suffer specific relations of exploitation, unequal bargaining and discrimination within the political economy, and there may also be historical reasons why their homes and sources of livelihood are located in resource-poor areas.[50]

The relationship between discrimination and exposure and vulnerability to disaster-related harm is a significant one that will be revisited in Chapter 6. For present purposes, it is sufficient to highlight that the connection is well established in the social sciences literature, as confirmed by Twigg:

---

[47] Oliver-Smith, 'What Is a Disaster?' (n 27) 30.   [48] Bankoff, *Cultures of Disaster* (n 25) 157–58.
[49] John Twigg, *Disaster Risk Reduction: Mitigation and Preparedness in Development and Emergency Programming*, ODI Humanitarian Practice Network Good Practice Review No. 9 (March 2004) 248. <https://www.preventionweb.net/educational/view/8450> accessed 4 April 2019.
[50] Wisner et al, *At Risk* (n 32) 56.

Extensive research over the past 30 years has shown that, in general, it is the weaker groups in society that suffer worst from disasters: the poor (especially), the very young and the very old, women, the disabled, and those who are marginalised by race or caste ... Those who are already at an economic or social disadvantage tend to be more likely to suffer during disasters.[51]

This observation is clearly relevant to RSD in the context of 'natural' disasters and climate change.

2.3.3 Definition

Hence, for the purposes of this book, and having regard to the specific considerations relevant to the individual nature of refugee status determination, but also addressing calls elsewhere for recognition of smaller-scale 'natural' disasters within formal definitions,[52] a 'natural' disaster is defined in this book as:

> A situation in which the accrual of one or more natural hazard events or processes of environmental change (such as cyclones, droughts, earthquakes, sea level rise etc.) interacts with vulnerable and exposed social conditions resulting in seriously adverse human impacts on a scale that affects more than just one household.

Identification of a few key features of the definition will suffice at this juncture. First, 'natural' disasters result from the interaction of natural hazards with exposed and vulnerable social conditions. Both natural hazards and social conditions are preconditions to the unfolding of a disaster. Second, 'natural' disasters, as conceptualised in the working definition, can occur suddenly but may also unfold over time. The potentially broad temporal scope of the concept reflects key insights from disaster anthropology and political ecology developed in this chapter. Third, 'natural' disasters have serious adverse impacts. The occurrence of natural hazard events is not exceptional, and societies have developed ways of living with risk. Thus, floods that lead to the closure of a bridge, thereby preventing traffic from crossing it for a week, may or may not have seriously adverse human impacts. Adverse human impacts might include reduced access to food, water, shelter, livelihoods, and healthcare. Fourth, and related to the third point about human impacts, is the inherent need to consider the specific local context. A bridge closure may not amount to a disaster if there is another bridge in operation nearby or other means of transporting essential goods and services between the two points, though it could have serious adverse human impacts where households find themselves entirely cut off. Finally, the working definition is alive to questions of scale, which tend not to be reflected in dominant definitions of disaster. By focusing on events or processes affecting more than one household, the definition avoids undue emphasis on unitary notions such as 'community' or 'society' that detract focus from differential exposure and vulnerability. The definition recognises the existence of small-scale disasters that may not, as dominant definitions require, overwhelm the capacity of the state to respond.[53]

---

[51] Twigg, 'Disaster Risk Reduction' (n 49) 16.
[52] See for example Wisner et al, 'Framing Disaster' (n 43) 30.   [53] See discussion in Section 2.1.

2.3.4 Discrimination

The preceding section presented the social paradigm of disasters, which emphasises how historical and contemporary social processes interact with environmental factors to engender vulnerability and exposure to disaster-related harm. The social paradigm's recognition that some people may be more exposed and vulnerable to disaster-related harm than others was also considered to represent a significant contribution to the understanding of 'natural' disasters.

This section now sets out to expand upon the notion of differential exposure and vulnerability to disaster-related harm by focusing specifically on the relevance of discrimination in that connection.

The PAR model, and wider approaches under the umbrella of the 'social' paradigm described above, recognises that particular groups within society can be particularly vulnerable to disaster-related harm. Indeed, the assumption reflected in the hazard paradigm that disasters impact indiscriminately is roundly rejected under the social paradigm. Wisner et al explain:

> people's exposure to risk differs according to their class (which affects their income, how they live and where), whether they are male or female, what their ethnicity is, what age group they belong to, whether they are disabled or not, their immigration status, and so forth.[54]

They continue:

> Many aspects of the social environment are easily recognised: people live in adverse economic situations that oblige them to inhabit regions and places that are affected by natural hazards, be they the flood plains of rivers, the slopes of volcanoes or earthquake zones. However, there are many other less obvious political and economic factors that underlie the impact of hazards. These involve the manner in which assets, income and access to other resources, such as knowledge and information, are distributed between different social groups, and various forms of discrimination that occur in the allocation of welfare and social protection (including relief and resources for recovery). It is these elements that link our analysis of disasters that are supposedly caused mainly by natural hazards to broader patterns in society.[55]

Drawing focus to these 'less obvious political and economic factors', including discrimination, the social paradigm reflects a concern with the wider context that engenders an individual's predicament. Its affinity with the predicament approach to RSD, discussed later in Chapter 3, is unmistakeable. Social processes, including discrimination, engender differential exposure and vulnerability to the kinds of harm that manifest in the context of 'natural' disasters.

The role of discrimination in engendering differential vulnerability and exposure is highlighted by social scientists working within the social paradigm. Havidán Rodríguez, for example, emphasises the continuity between existing social and economic conditions and apparently isolated disaster events:

> Disasters are human-induced, socially constructed events that are part of the social processes that characterize societies throughout the world. Rather than isolated events,

---

[54] Wisner et al, *At Risk* (n 32) 6. [55] Ibid 5.

disasters should be viewed and studied as part of the normal fabric of societies, which reflect their social, political and economic structures and social organization (or lack thereof). Furthermore, disasters are not random or equal probability events but are the result of existing social and economic conditions. Therefore, it should come as no surprise that those disenfranchised from political and economic power disproportionately suffer the consequences of these events and have the greatest difficulties in recovering from them. In fact, disasters serve to bring to the forefront the social inequities that characterize contemporary societies.[56]

Discrimination and other structural factors can both engender and exacerbate differential exposure and vulnerability both to the impacts of the hazard event itself as well as to the challenges people face during and in the aftermath of such an event. The enduring character of discrimination may connect conduct over time to harm in particular instances.

The decidedly differential impact of at least some 'natural' disasters, understood from within the social paradigm, has also gained mainstream acceptance within the disaster risk reduction community. The International Federation of Red Cross and Red Crescent Societies (IFRC), for example, devoted the 2007 edition of its annual World Disasters Report to the subject. Providing numerous examples of discrimination in the context of disasters, this publication emphasised that:

> Discrimination was and is inherent in many societies, with disasters often magnifying the problem ... Every emergency involves people who cannot access food and shelter simply because of their age, ethnicity, gender or disability. People already on the margins of society as a result of discrimination are made even more vulnerable through a crisis.[57]

So not only are people who experience discrimination in everyday life more exposed and vulnerable to disaster-related harm, but the impact of 'natural' disasters exacerbates this condition.

The Inter-Agency Standing Committee notes that '[e]xperience has also shown that preexisting vulnerabilities and patterns of discrimination usually become exacerbated in situations of natural disasters'.[58]

The Inter-Governmental Panel on Climate Change recognises the role of discrimination in determining vulnerability and exposure:

> People who are socially, economically, culturally, politically, institutionally, or otherwise marginalized are especially vulnerable to [the adverse impacts of] climate change ... This heightened vulnerability ... is the product of intersecting social processes that result in inequalities in socioeconomic status and income, as well as in exposure. Such social processes include, for example, discrimination on the basis of gender, class, ethnicity, age and (dis)ability.[59]

---

[56] Havidán Rodríguez and John Barnshaw, 'The Social Construction of Disasters: From Heat Waves to Worst-Case Scenarios' (2006) 35 Contemp Sociol 218, 222.
[57] IFRC, *World Disasters Report 2007: Focus on Discrimination* (2007) 12.
[58] IASC, 'IASC Operational Guidelines on the Protection of Persons in Situations of Natural Disasters' (The Brookings-Bern Project on Internal Displacement 2011) 2.
[59] IPCC, 'Summary for Policymakers' in *Climate Change 2014: Impacts, Adaptation, and Vulnerability. Part A: Global and Sectoral Aspects. Contribution of Working Group II to the Fifth Assessment Report of the Intergovernmental Panel on Climate Change* (CUP 2014) 6.

Similarly, the Nansen Initiative on Disaster-Induced Cross-Border Displacement recognises that:

> Because disasters exacerbate pre-existing social vulnerabilities and inequalities, children, women, older persons, persons with disabilities, impoverished communities, indigenous people, migrants and marginalized groups are more likely to have particular protection needs.[60]

What is beginning to emerge from this survey of scholarly and practitioner literature is an authoritative consensus that discrimination both engenders and exacerbates exposure and vulnerability to disaster-related harm. At this point, the difference between the hazard paradigm and the social paradigm is very clear, and the relevance of the latter to determining refugee status in the context of 'natural' disasters is inescapable. If, for reasons of race, religion, nationality, membership of a particular social group, or political opinion, a person is exposed and vulnerable to serious harm, the possibility of establishing a well-founded fear of being persecuted for a Convention reason in this connection is tangible, if not yet entirely clear.

Having recognised *that* discrimination plays a role in engendering differential vulnerability and exposure to disaster-related harm, the next question to be put is *how* this happens.

Wisner et al refer to studies that demonstrate the relevance of discrimination in engendering exposure and vulnerability to disaster risk:

> Ethnic divisions are often superimposed on class patterns, and may become the dominant factor determining vulnerability. This can be seen in differential access to, or possession of, resources, or inequalities of participation in different livelihoods, according to imposed racial or ethnic distinctions. However, very few studies of flood disasters seem to take up the issue of ethnicity as a vulnerability factor. One very significant case is that of the Venezuela floods/landslides of 1999 (see above), when Afro-Venezuelans were disproportionately affected. Another example comes from the exceptional flooding around Alice Springs in central Australia in 1985. Aboriginal people did not receive flood warnings and lived in flimsy accommodation on low-lying land. The radio broadcasts that alerted the white people were not on channels which were customarily used by the Aborigines.[61]

Gaillard echoes Serje's observation that '[i]n the same way that the study of hazards has dominated knowledge generation about disasters, the study of physical vulnerability has dominated the much less formal study of disaster vulnerability',[62] confirming the limited number of studies focusing on discrimination and disaster-related adversity:

> Ethnicity, caste and religious affiliations are most often neglected by those collecting and aggregating disaster-related data. Therefore, there is no extensive and useful dataset pertaining to the disaggregated impact of disasters on different ethnic groups, castes and religious communities on a global scale. Available evidence only reflects the scope of

---

[60] Nansen Initiative, 'Agenda for the Protection of Cross-Border Displaced Persons in the Context of Disasters and Climate Change' (2015) 6 <https://disasterdisplacement.org/the-platform/our-response> accessed 4 April 2019.
[61] Wisner et al, *At Risk* (n 32) 238.
[62] Julio Serje, 'Data Sources on Hazards' in Wisner et al (eds), *Handbook of Hazards* (n 41) 186.

research conducted in the field, which has so far been mostly in the USA (e.g. Bolin 2007; Bolin and Bolton 1986).[63]

Gaillard's call for a dataset pertaining to the disaggregated impact of disasters is echoed by both the UN Special Rapporteur on the Human Rights of Internally Displaced Persons[64] and the UN Human Rights Council.[65] Bolin considers that '[t]his is an area in the disaster literature where there is clearly a need for more place-specific, historically informed case studies'.[66]

However, although limitations in the kind of 'place-specific, historically informed case studies' or even more generalised 'disaggregated data' leave room for nuanced development and potential refinement of the social paradigm, a range of specific examples provide insight into the kinds of circumstances in which discrimination may engender and exacerbate exposure and vulnerability to disaster-related harm. In what follows, a compilation of examples from different 'natural' disaster contexts provides insight into *how* discrimination engenders differential exposure and vulnerability to disaster-related harm. Impacts of discrimination can manifest before, during, and in the aftermath of the unfolding of a 'natural' disaster situation.

Twigg identifies discrimination that engenders differential exposure and vulnerability to disaster-related harm:

> The exclusion and attendant poverty of ethnic minorities may force them into settlement in dangerous locations, or to live on land of poor quality that produces little food, while language, educational and cultural barriers can restrict access to information on risk and risk avoidance. Migrants can be doubly vulnerable: as members of minority ethnic groups, they may be neglected or even persecuted; as strangers to an area they lack the knowledge and coping strategies to protect themselves.[67]

Reflecting on their study of human rights in countries affected by the 2004 Indian Ocean tsunami, Fletcher et al explain:

> Natural disasters can exacerbate pre-existing vulnerabilities of populations already at risk. Poverty-stricken groups living in substandard housing, on unstable ground, or in flood plains are usually the principal victims of these disasters. Often these groups have experienced ongoing discrimination because of their ethnicity, religion, class, or gender, which has left them living in fragile physical environments ... In countries where

---

[63] JC Gaillard, 'Caste, Ethnicity, Religious Affiliation and Disaster' in Wisner et al (eds), *Handbook of Hazards* (n 41) 460.
[64] UNHRC, *Report of the Special Rapporteur on the Human Rights of Internally Displaced Persons* (29 April 2016) A/HRC/32/35 [77]: 'Greater research and data is required globally to reveal the full impact of displacement on such communities, as well as regional trends, patterns and dynamics of displacement. In particular, this makes it necessary to disaggregate data not only by sex and age but also by diversity categories, such as ethnicity and religion, that should be determined by contextual realities.'
[65] UNHRC, 'Human Rights of Internally Displaced Persons' (29 June 2012) A/HRC/20/L.14: 'the importance of the effective collection of data, disaggregated by age, sex, diversity and location, on internally displaced persons for the protection of their human rights, the implementation of durable solutions and the assessment of their specific needs and vulnerabilities, and encourages Governments to use, on a voluntary basis, the services of the Inter-Agency Joint Internally Displaced Person Profiling Service, which has been set up to offer technical support in this regard.'
[66] Bob Bolin 'Race, Class, Ethnicity and Disaster Vulnerability' in H Rodríguez, EL Quarantelli and RR Dynes (eds), *Handbook of Disaster Research* (Springer 2007) 123.
[67] Twigg, 'Disaster Risk Reduction' (n 49) 99.

corruption and bureaucratic incompetence are rife, certain individuals and groups may manipulate their political connections to receive or distribute aid at the expense of others. Still other groups may receive little or no aid because of their ethnicity, religion, gender, age, or social standing.[68]

This evidence is corroborated by Gill, based on fieldwork in India in the aftermath of the tsunami:

> Dalit communities were more vulnerable than other groups to the disaster before it happened due to pre-existing debts, low savings, poor quality settlements, lack of assets, low social status, dirth [sic] of social capital in the form of effective organisations and ability to 'plug into' media and social networks, lack of effective political representation, and their reliance on daily wage labour. Despite the lesser loss of life and property, in many ways Dalits have suffered more greatly than other groups as a result of their comparative vulnerability, poverty and invisibility.[69]

Such a relationship between pre-existing patterns of discrimination and differential exposure and vulnerability to disaster-related harm is not limited to the Indian Ocean tsunami. Söderbergh identifies the operation of both aspects of the differential impact of 'natural' disasters in his reference to flooding in India in 2007:

> Dalits, Adivasis and Muslims were especially vulnerable to harm during the unusually severe monsoon floods in India of 2007. Their homes were more prone to damage, and these communities were often the last to receive assistance as aid workers were unaware of their marginal locations or because majority population representatives took charge of the distribution of relief. Thus, in surveys following the floods, Dalits represented the overwhelming majority of fatalities.[70]

The Special Rapporteur on the Human Rights of Internally Displaced Persons explains:

> Vulnerability to displacement may be heightened by discriminatory State policies or practices. Non-documentation, the denial or deprivation of citizenship for some ethnic or religious groups, for example, renders them stateless. Their rights as citizens are not fully recognized and they may be targeted, or not adequately protected, by national authorities.[71]

A classic example is provided by what has been described as the 'class-quake' in Guatemala in 1976. Indigenous Maya communities living on steep mountain slopes in cheap adobe or other shelters comprised a disproportionate number of fatalities. Wisner et al describe the differential impact from a social perspective:

---

[68] Laurel E. Fletcher, Eric Stover and Harvey M. Weinstein, 'After the Tsunami: Human Rights of Vulnerable Populations' (Human Rights Center of UC Berkeley and East-West Center 2005) 1.

[69] Timothy Gill, 'Making Things Worse: How "Caste-Blindness" in Indian Post-Tsunami Disaster Recovery Has Exacerbated Vulnerability and Exclusion' (2007) Dalit Network Netherlands, 12 <http://idsn.org/uploads/media/Making_Things_Worse_report.pdf> accessed 4 April 2019.

[70] Carl Söderbergh, 'Human Rights in a Warmer World: The Case of Climate Change Displacement' (2011) Working Paper 2011-01-28, 17–18 <http://lup.lub.lu.se/record/1774900> accessed 4 April 2019, citing Rachel Baird, 'The Impact of Climate Change on Minorities and Indigenous Peoples' (2008) Minority Rights Group International Briefing <http://minorityrights.org/publications/the-impact-of-climate-change-on-minorities-and-indigenous-peoples-april-2008/> accessed 4 April 2019.

[71] UNHRC, *Report of the Special Rapporteur* (n 64) [79].

slum dwellers in Guatemala City and many Mayan Indians living in impoverished towns and hamlets suffered the highest mortality. The homes of the middle class were better protected and more safely sited, and recovery was easier for them. The Guatemalan poor were caught up in a vicious circle in which lack of access to means of social and self-protection made them more vulnerable to the next disaster. The social component was so apparent that a journalist called the event a 'class-quake'.[72]

From the above examples, some of the numerous ways in which discrimination engenders and exacerbates differential exposure and vulnerability to disaster-related harm include:

- Differential access to resources
- Inequalities in participation in different livelihoods
- Differential access to information
- Poor quality shelter
- Shelter in dangerous locations
- Debt
- Low savings
- Lack of assets
- Low social capital
- Lack of political representation
- Non-documentation
- Statelessness
- Elite capture
- Exclusion from relief

Hence, it is clear that there are numerous ways in which *ex ante* discrimination engenders exposure and vulnerability to serious disaster-related harm, and this is directly relevant to approaching claims for recognition of refugee status.

An important note of caution needs to be made here, however. People may be exposed and vulnerable, and discrimination may be a significant factor contributing to such a predicament. However, reflecting the deeper debate within the social sciences between structure and agency,[73] it warrants recalling, both for theoretical purposes as well as for the specific purpose of RSD, that individuals still make choices based on the options available to them. Their options may well be circumscribed as a consequence of discrimination, but it would be overstating the role of structure to suggest that all people who, partly as a consequence of discrimination, inhabit a flood plain or poorly irrigated slum settlement are there solely as a consequence of the social forces that constrain them. Flood plains, for example, tend to provide rich soil, inviting a trade-off between the risk of flooding and the gains to be made from better quality land.[74] One critique of the PAR model is its emphasis on structural factors.[75]

---

[72] Wisner et al, *At Risk* (n 32) 9.
[73] To explore this protracted debate, consider the four-volume set Mike O'Donnell (ed), *Structure and Agency* (Sage 2010).
[74] Wisner et al, *At Risk* (n 32) 202.
[75] Keith Smith, *Environmental Hazards: Assessing Risk and Reducing Disasters* (2nd edn, Routledge 1996), cited in Wisner et al, *At Risk* (n 32) 30.

The limited focus attached to individual agency under the PAR model notwithstanding, from the foregoing it is clear that the social paradigm provides an indispensable theoretical framework for approaching RSD in the context of 'natural' disasters.

2.3.5 Structural Violence

Differential exposure and vulnerability to disaster-related harm is not isolated from the wider social context. As highlighted above, it results from the direct as well as indirect conduct of state and non-state actors. It results from decisions taken by the state to suppress certain political ideas or the expression of certain religious beliefs. It is identifiable in the social environment that reproduces negative attitudes towards people with non-heteronormative sexual orientations and/or gender identities. It is identifiable in everyday racism and patriarchy. This discriminatory social and political environment plays out, at times across an entire lifespan or even generations.

In many cases, it will be readily apparent who the actor whose discriminatory conduct contributes to differential exposure and vulnerability to disaster-related harm is. In other cases, the discriminatory conduct will be more diffuse. For example, private actors, operating within a discriminatory social environment that makes unjust distinctions between men and women, contribute to the differential exposure and vulnerability of women to disaster-related harm. The diffuse nature of the conduct reflected in indirect and systemic discrimination does nothing to mitigate the seriousness of harm to which an individual may be exposed. Relevant actors in this connection are those who: deny access to education; restrict access to livelihood activities; fail to provide the highest attainable standard of health, adequate food, shelter, and so forth. They are those who enforce social codes relegating minority ethnic or caste groups to unsafe living conditions in exposed locations. They are the state and non-state actors who enact, condone, or fail to condemn discriminatory practices.

The above propositions find support in the notion of structural violence, which represents an academic attempt to conceptualise the kind of violence reflected in, amongst other practices, systemic discrimination. First articulated by Johan Galtung in his seminal article of 1969,[76] the notion of structural violence recognises the violence inherent in discrimination, and also emphasises that the temporal scope of violence is not delimited to concrete instances of cause and effect, such as the application of an electric shock to a victim of torture, or the unlawful arrest and detention of a political dissident. Structural violence and more immediate forms of physical, interpersonal violence are not, however, isolated phenomena, but rather 'cross-breed',[77] much like the relationship between being persecuted and being the victim of an act of persecution, as discussed in Chapters 5 and 6.

---

[76] Johan Galtung, 'Violence, Peace, and Peace Research' (1969) 6 J Peace Res 167. It is important to emphasise their focus on macro-level processes, including the global, neo-liberal economic order, colonialism and post-colonialism, climate change, and so forth. These macro-level forces are of course relevant to individual predicaments as well, but they do not form part of the conceptualisation of disasters that informs RSD. The features of structural violence and slow violence that do inform this conceptualisation are the recognition that violence may unfold across a broad temporal landscape and can be enacted by a number of actors across time and space. In a sense, it is thus the local and national facets of structural and slow violence that can inform RSD in the context of disasters and climate change.

[77] Ibid 178.

Appreciation of structural violence is evident within the international human rights machinery. For example, the Report of the Special Rapporteur on Violence against Women, its Causes and Consequences explains:

> Institutional and structural violence is any form of structural inequality or institutional discrimination that maintains a woman in a subordinate position, whether physical or ideological, to other people within her family, household or community. In many contexts, there are discriminatory measures in place that maintain gender stratification that privileges male power and control, and which disadvantages some women in particular ways.[78]

Importantly, the Special Rapporteur emphasises that violence is to be understood with reference to a continuum between structural and interpersonal manifestations, without hierarchy. She explains:

> Situating violence along a continuum allows for an appropriate contextualizing of violence in that the deprivation of water, food, and other human rights can be just as egregious and debilitating as family violence. Although these forms of violence are by no means the same, they can be viewed as parallel and similar when considering their interrelationship.[79]

The Special Rapporteur identifies a range of practices and beliefs that fall within the ambit of structural violence, including:

- Laws and policies that maintain one group's advantage over another in places of employment, and in terms of educational opportunities, access to resources, forms and places of worship, protection by the police and other state forces, and government services and benefits
- An absence of laws that criminalise all forms of violence against women
- Differential inheritance, land tenure, and property ownership practices
- Beliefs that perpetuate the notion that males are superior to females, that whites are superior to blacks, that persons without physical or mental impairment are superior to those with disabilities, that one language is superior to another, and that one class position is entitled to rights denied to another, are all factors contributing to structural violence that have become institutionalised forms of multiple and intersecting discrimination
- Gender ideologies that dictate that men should control women or allow for men to physically control their partners or children

Finally, the report is founded upon an appreciation of the compounding effect of multiple discrimination, for example when gender, race, disability, and other factors intersect in an individual or group predicament.

Similar examples are to be found in the context of discrimination on the grounds of race, religion, nationality, age, and disability.[80]

---

[78] Rashida Manjoo, *Report of the Special Rapporteur on Violence against Women, Its Causes and Consequences* (2 May 2011) A/HRC/17/26, [26].
[79] Ibid [66].
[80] For other examples see UNHRC, *Report of the Commission on Human Rights in South Sudan* (6 March 2017) A/HRC/34/63; General Assembly, Security Council, *Causes of Conflict and the Promotion of Durable Peace and Sustainable Development in Africa: Report of the Secretary General* (26 July 2016)

Having established that structural violence is a concept that is acknowledged within the international human rights machinery, elements of the concept as originally articulated and developed over time warrant consideration to the extent that they add further insight relevant to the determination of refugee status.

Galtung's theory is summarised below:

> We shall refer to the type of violence where there is an actor that commits the violence as personal or direct, and to violence where there is no such actor as structural or indirect. In both cases individuals may be killed or mutilated, hit or hurt in both senses of these words, and manipulated by means of stick or carrot strategies. But whereas in the first case these consequences can be traced back to concrete persons as actors, in the second case this is no longer meaningful. There may not be any person who directly harms another person in the structure. The violence is built into the structure and shows up as unequal power and consequently as unequal life chances.[81]

Galtung's notion of 'structural violence' provides a language for recognising human agency in situations that might otherwise be described as 'adversity' and helps to show how cases that may at first appear as arising from the indiscriminate forces of nature should instead be understood as failures of state protection.

Paul Farmer has applied the notion of structural violence to the HIV/AIDS epidemic in Haiti. Adopting a similar perspective to that used by Oliver-Smith in his description of Haiti's 500-year earthquake,[82] Farmer, recognising Galtung, defines structural violence as follows:

> Structural violence is violence exerted systematically – that is, indirectly – by everyone who belongs to a certain social order: hence the discomfort these ideas provoke in a moral economy still geared to pinning praise or blame on individual actors. In short, the concept of structural violence is intended to inform the study of the social machinery of oppression. Oppression is a result of many conditions, not the least of which reside in consciousness. We will therefore need to examine, as well, the roles played by the erasure of historical memory and other forms of desocialization as enabling conditions of structures that are both 'sinful' and ostensibly 'nobody's fault'.[83]

Farmer's definition and description highlight much of the promise of structural violence as a concept linking the social paradigm of disaster studies to RSD. It highlights how the concept identifies violence that is enacted 'indirectly' and in ways not delimited to the conduct of specific individual actors. It recognises how the apparently diffuse nature of this

---

A/71/211–S/2016/655; CRPD, 'General Comment No. 3: Article 6: Women and Girls with Disabilities' (2 September 2016) CRPD/C/GC/3; UNHRC, *Report of the Special Rapporteur on the Rights of Indigenous Peoples, Victoria Tauli Corpuz* (6 August 2015) A/HRC/30/41; UNHRC, *Report of the Special Rapporteur on Freedom of Religion or Belief, Heiner Bielefeldt* (29 December 2014) A/HRC/28/66; UNHRC, *Annual Report of the Special Representative of the Secretary-General on Violence against Children* (30 December 2014) A/HRC/28/55; UNHRC, *The Right of Peoples to Peace: Progress Report Prepared by the Drafting Group of the Advisory Committee* (9 December 2011) A/HRC/AC/8/2; UNHRC, 'Accelerating Efforts to Eliminate all Forms of Violence against Women: Ensuring Due Diligence in Prevention' (16 June 2010) A/HRC/14/L.9/Rev.1.

[81] Galtung, 'Violence, Peace, and Peace Research' (n 76) 171.
[82] Oliver-Smith, 'Haiti's 500-Year Earthquake' (n 44).
[83] Paul Farmer, 'An Anthropology of Structural Violence' (2004) 45 Curr Anthropol 305, 307.

violence enables the impacts to be described as 'nobody's fault', even whilst the conduct of whole social strata may be implicated.

It is possible to place too much emphasis on the relevance of the concept of structural violence to the determination of refugee status in the context of 'natural' disasters, however. Despite mentioning race, Farmer does not focus on discrimination on an individual or group level in his analysis. In Farmer's analysis of structural violence in Haiti, the relevant conduct includes the construction of a slave economy by the French, the destruction and death caused during the rebellion, the imposition of a duty to compensate France for this loss of its colony, the maintenance of an embargo against the new republic, and successive waves of US military intervention in the economic and political life of the country. Notwithstanding his macro-level approach, Farmer's appreciation of how this structural violence delimits individual capacities is insightful:

> Structural violence is structured and *structuring*. It constricts the agency of its victims. It tightens a physical noose around their necks, and this garrotting determines the way in which resources – food, medicine, even affection – are allocated and experienced.[84]

The concept of structural violence is also effective in its express recognition of the expanded temporal scope of such conduct. Although clearly more diffuse than the intentional application of electrodes to the body in an act of torture, the slow violence[85] that entails limits on access to education, healthcare, adequate shelter, work, property, food, information, participation in public life, and so forth, which together delimit the range of an individual's capabilities and thereby engender exposure and vulnerability to a range of shocks, including 'natural' disasters, is as relevant to RSD. Combine these 'root causes' with dynamic threats such as discriminatory failure to maintain flood defences in slum areas, preferential placing of storm shelters in proximity to power, dominance of certain social groups combined with an inability on the part of the state to protect marginalised populations, for example in situations of acute food insecurity, and so forth, and this 'slow violence' results in unsafe conditions for marginalised individuals. Understanding discrimination as structural violence, even when the effects of such conduct are delayed, reveals how discrimination in everyday life may engender or exacerbate a real chance of exposure to serious harm. In this way, serious denials of human rights in 'natural' disaster situations that appear as 'nobody's fault' are instead understood as reflecting a condition of existence in which discrimination is a contributory cause of differential exposure and vulnerability to disaster-related harm. This insight is critical for an appreciation of the kinds of circumstances in which a person may establish eligibility for recognition of refugee status in the context of disasters and climate change. As the next chapter demonstrates, the perspective has gained some judicial recognition, but is far from overtaking the dominant view described in Chapter 1.

---

[84] Ibid 308.
[85] Slow violence is a concept that builds on notions of structural violence, but emphasises the ways in which historical conduct has contemporary impacts. See Rob Nixon, *Slow Violence and the Environmentalism of the Poor* (Harvard University Press 2013).

3

Jurisprudence on the Determination of Refugee Status in the Context of 'Natural' Disasters and Climate Change

3.1 Methodology

An extensive review was conducted of judicial decisions on claims for recognition of refugee status in the context of 'natural' disasters and climate change. In order to analyse a wide range of disaster-related scenarios in which refugee status determination (RSD) has been performed, a thorough review of jurisprudence available on the UNHCR's RefWorld database was conducted. Case law is predominantly in English, although material in French, Spanish, German, and Portuguese is also said to be available.[1]

Relevant cases were identified by using the 'advanced search' function and searching for the following terms: climate change; cyclone; disaster; drought; earthquake; environmental degradation; famine; flood; food aid; hurricane; landslide; sea level; storm; and tsunami. The search was conducted only in English, which is a limitation. However, after accounting for duplicate references, cases including more than one search term, and cases from the international criminal courts and tribunals,[2] the search nevertheless returned approximately 500 cases from the following jurisdictions: Australia, Canada, China (Hong Kong), Ecuador, India, Ireland, Israel, Italy, Japan, Kenya, Malawi, Moldova, New Zealand, Norway, the Russian Federation, South Africa, the United Kingdom, and the United States of America, as well as regional and international judicial bodies, including the African Court of Human and Peoples' Rights, the European Court of Human Rights, the Inter-American Court of Human Rights, the International Court of Justice, and the Human Rights Committee of the ICCPR. A preliminary filtering of results narrowed the relevant or potentially relevant number of cases down to approximately 200.

Cases identified as being relevant included those where an express connection was made, either by the claimant or by the judicial body, between hazard events, disasters, and/or climate change, and the basis of claim. The identification of relevant cases was generous in order not to prematurely exclude cases of potential relevance. Cases where state practice was referenced,[3] cases where general comments were made about a relationship between

---

[1] UNHCR, 'UNHCR's Refworld Case Law Collection User Guide, Status Determination and Protection Information Section (SDPIS)' (January 2009) <http://www.refworld.org/docid/497da4f82.html> accessed 4 April 2019.

[2] Cases from international criminal courts and tribunals were not considered as part of the case law review because, although there may conceivably be isolated incidents in which an individual's role in a disaster could engage criminal liability under international law (for example, in the context of denial of food aid), the focus of these judicial bodies was considered to lie outside the scope of this research.

[3] *Matter of Echeverria*, 25 I&N Dec. 512 (BIA 2011), United States Board of Immigration Appeals, 1 June 2011.

hazards, disasters, or climate change and migration,[4] and cases where specific country conditions were identified, even if in passing,[5] were all considered relevant. Thus, a case in which an Afghan national had expressed fear of being harmed by the Taliban was retained because of reference to recurrent disasters in the country.[6]

Cases were identified as being irrelevant where a search term was used in a manner that was irrelevant to the study. For example, where the term 'disaster' was used to relate to adverse personal outcomes, such as a 'business disaster'[7] or social outcomes not relating to natural hazards, for example, where it was said that 'the elections in October 2000 were a disaster',[8] the case was not deemed relevant and as such was excluded.

Having reduced the number of relevant cases to approximately 200, it was possible to formulate a rough taxonomy, which is presented in Section 3.3, following a brief overview of the key elements of the refugee definition. This brief overview is a necessary prerequisite for the subsequent analysis of jurisprudence, as key tensions that pervade international refugee law doctrine are clearly evident in the jurisprudence relating to RSD in the context of 'natural' disasters and climate change.

3.2 Basic Principles of International Refugee Law

Under Article 1A(2) of the Refugee Convention, a refugee is a person who,

> owing to a well-founded fear of being persecuted for reasons of race, religion, nationality, membership of a particular social group or political opinion, is outside the country of his nationality and is unable or, owing to such fear, is unwilling to avail himself of the protection of that country; or who, not having a nationality and being outside the country of his former habitual residence as a result of such events, is unable or, owing to such fear, is unwilling to return to it.

In order to be recognised as a refugee under the Refugee Convention, a person must establish that she is a person to whom the definition at Article 1A(2) applies. Each element of the definition must be satisfied in order for the person to be recognised as a refugee. The core elements of the refugee definition require the claimant, once outside the country of her nationality or former habitual residence, to establish (1) a well-founded fear (2) of being persecuted (3) for reasons of (4) race, religion, nationality, membership of a particular social group, or political opinion.[9] The core elements are to be approached holistically,[10]

---

[4] *Advisory Opinion OC-21/14 of August 19, 2014 requested by the Argentine Republic, the Federative Republic of Brazil, the Republic of Paraguay and the Oriental Republic of Uruguay: Rights and Guarantees of Children in the Context of Migration and/or In Need of International Protection*, OC-21/14, Inter-American Court of Human Rights (IACrtHR), 19 August 2014.
[5] *RRT Case No. 1213171* [2013] RRTA 157 (15 February 2013). [6] Ibid
[7] *SZOGB v Minister for Immigration and Another* [2010] FMCA 748, [66].
[8] *Alfarsy v Canada (Minister of Citizenship and Immigration)* [2003] FCJ No. 1856; 2003 FC 1461, [24].
[9] See UNHCR, 'Interpreting Article 1 of the 1951 Convention Relating to the Status of Refugees' (April 2001) <http://www.refworld.org/docid/3b20a3914.html> accessed 4 April 2019, 2: 'The key to the characterisation of a person as a refugee is risk of persecution for a Convention reason'.
[10] *Applicant A v Minister for Immigration and Ethnic Affairs* [1997] HCA 4, [1997] 190 CLR 225, 254 (McHugh J) and *Karanakaran v Secretary of State for the Home Department* [2000] EWCA Civ. 11, [2000] Imm AR 271, [19] (Sedley LJ).

although specific interpretations of each of the four elements have been developed over time, albeit not in a uniform manner across all jurisdictions where RSD is conducted.

In what follows, a very simplified overview of the four key elements of the refugee definition is provided. The main purpose is to put the remainder of the chapter in context by describing briefly what each element is generally considered to require. A much more detailed treatment of the refugee definition is conducted in Chapters 4–6, which is also informed by this subsection.

3.2.1 Well-Founded Fear

As to the first element – 'well-founded fear' – academic and judicial authorities reflect a general consensus that the term has the effect of directing decision-makers towards an assessment of the likelihood that a person will be persecuted for one or more of the five Convention reasons on return to her country of origin or habitual residence. There is consensus that the terms 'real chance', 'reasonable likelihood', and 'real risk' reflect a standard of proof lower than the civil balance of probabilities standard.[11] Although reference to statistical probability has been discouraged,[12] there is also recognition that a well-founded fear could be established where there was a 10 percent chance of a risk accruing, bringing the threshold well below the balance of probabilities.[13] There remains a degree of uncertainty about the role of a person's subjective fear in the assessment, with UNHCR guidelines[14] and some leading judgments[15] considering the subjective element important, whilst other judgments and commentary from legal doctrinal scholars seeing little or no room for consideration of subjectivity in what is essentially an objective, forward-looking assessment of conditions in the claimant's home country.[16] Agreeing with Hathaway that 'fear' should not be understood as necessitating a feeling of trepidation,[17] but rejecting any notion of a subjective–objective dichotomy, Noll articulates a persuasive

---

[11] See leading cases from the USA, United Kingdom, and Australia respectively: *INS v Cardoza-Fonseca*, 480 US 421 [1987]; *R v Secretary of State for the Home Department, ex parte Sivakumaran and Conjoined Appeals (UN High Commissioner for Refugees Intervening)* [1987] UKHL 1, [1988] AC 958; *Chan v Minister for Immigration and Ethnic Affairs* [1989] HCA 62, [1989] 169 CLR 379.

[12] See for example *S v Chief Executive of the Department of Labour and Another* [2005] NZHC 439 HC AK CIV 2005-404-003360.

[13] See for example the judgment of Stevens J in *Cardoza-Fonseca* (n 11) and approved of by Lord Keith in *Sivakumaran* (n 11) 994.

[14] Contrast UNHCR, *Handbook and Guidelines on Procedures and Criteria for Determining Refugee Status under the 1951 Convention and the 1967 Protocol Relating to the Status of Refugees* (reissued December 2011) <http://www.unhcr.org/publications/legal/3d58e13b4/handbook-procedures-criteria-determining-refugee-status-under-1951-convention.html> accessed 4 April 2019 [37–38] with Gregor Noll, 'Evidentiary Assessment under the Refugee Convention: Risk, Pain and the Intersubjectivity of Fear', in Gregor Noll (ed), *Proof, Evidentiary Assessment and Credibility in Asylum Procedures* (Brill 2005).

[15] *Cardoza-Fonseca* (n 11).

[16] For a clear rejection of the subjective approach, see James Hathaway and Michelle Foster, *The Law of Refugee Status* (2nd edn, CUP 2014) ch 2 and earlier James Hathaway and William Hicks, 'Is There a Subjective Element in the Refugee Convention's Requirement of "Well-Founded Fear"?' (2005) 26 Mich J Intl L 505. See also *Refugee Appeal No. 70074/96* RSAA (17 September 1996) and its discussion of the approach taken in leading authorities on the well-founded fear criterion, including *Sivakumaran* (n 11) and *Chan* (n 11).

[17] Noll, 'Intersubjectivity of Fear' (n 14) 143.

approach that understands the role of the term 'fear' as inviting consideration of the perspective of the claimant:

> What, then, could the meaning of 'fear' be? To our understanding, the occurrences of the term fear, as well as the explicit linkage between fear and unwillingness, suggest that refugee status determination under Article 1A(2) CSR involves the applicant's own assessment of her situation upon return.[18]

In other words, 'fear translates into a procedural standard'.[19]

Having heard the claimant and considered her own reasons why she is unable or unwilling to return owing to a fear of being persecuted for a Convention reason, the decision-maker will generally be required to seek information beyond the statements of the claimant. Consideration of additional information should not, however, be seen as a purely 'objective' exercise contrasted with the 'subjective' fear of the claimant, as:

> The total subjugation of fear to reason disenfranchises the refugee, while framing the Northern adjudicator as a gnostic agent, capable of better understanding reality in the South or the East through its Northern institutions.

This book adopts Noll's view that the 'well-founded fear' element of the refugee definition entails a procedural standard to hear the claimant and to assess the claim in light of all available evidence.

3.2.2 ... of Being Persecuted

The authorities are even less united in their interpretation of the 'being persecuted' element of the definition. Goodwin-Gill and McAdam, for example, consider that the lack of a definition of the term in the Convention itself is warranted, given that the concept is 'only too readily filled by the latest examples of one person's inhumanity to another',[20] echoing Paul Weis's early observation that 'the term "persecution" has nowhere been defined and this was probably deliberate'[21] and Grahl-Madsen's opinion that the drafters probably 'wanted to introduce a flexible concept which might be applied to circumstances as they might arise; or in other words, that they capitulated before the inventiveness of humanity to think up new ways of persecuting fellow men'.[22] It is recognised that, at the very least, a threat to life or freedom will amount to persecution, owing to the inclusion of these threats at Article 33 of the Convention.[23]

Hathaway and Foster[24] attempt a more systematic approach to the meaning of being persecuted, and their work, informed by Hathaway's first treatise on the law of refugee status,[25] is widely recognised as an authoritative statement on the human rights-based

---

[18] Ibid [19] Ibid 144.
[20] Guy S. Goodwin-Gill and Jane McAdam, *The Refugee in International Law* (3rd edn, OUP 2007) 93.
[21] Paul Weis, UN Doc. HCR/INF/49, 22 cited in Atle Grahl-Madsen, *The Status of Refugees in International Law* (AW Sijthoff's Uitgeversmaatschappij NV 1966) 193.
[22] Grahl-Madsen, *The Status of Refugees in International Law* (n 21) 193.
[23] UNHCR, *Handbook* (n 14) [51].   [24] Hathaway and Foster, *Refugee Status* (n 16).
[25] James Hathaway, *The Law of Refugee Status* (Butterworths 1991).

approach to the refugee definition.[26] On this approach, being persecuted is defined as 'a sustained or systemic denial of human rights demonstrative of a failure of state protection'.[27] The concept is understood with reference to those international human rights instruments ratified by a 'super-majority of states across a politically and geographically diverse range of states'.[28] Hence, when rights protected under the international bill of rights in particular (namely the UDHR,[29] ICCPR,[30] and ICESCR[31]), but also widely ratified instruments, such as the CRC,[32] are denied in a manner that is sufficiently serious, that may amount to being persecuted.

The UNHCR has expressed concern about the wholesale incorporation of international human rights standards into international refugee law, considering that, whilst violations of human rights may well be indicative of persecution, the term should not be rigidly defined:

> The interpretation of persecution needs to be flexible, adaptable and sufficiently open to accommodate ever changing forms of persecution and human rights abuses. Persecution cannot and should not be defined solely on the basis of serious human rights violations. Severe discrimination or the cumulative effect of various measures not in themselves alone amounting to persecution, as well as their combination with other adverse factors, can give rise to a well-founded fear of persecution.[33]

This variation notwithstanding, there is broad consensus that being persecuted entails being exposed to serious harm, including deprivation of civil and political as well as economic and social rights, from which the state is unwilling or unable to protect. The maxim adopted by Lord Hoffman in *Islam* reflects this 'bifurcated' approach:

Persecution = Serious Harm + the Failure of State Protection[34]

The need to establish not only that the person faces exposure to serious harm, but also that the state would be unable or unwilling to provide protection is intimately connected to contemporary understandings of the purpose of the Refugee Convention being to extend surrogate international protection when the claimant's home state cannot or will not do so.[35] That the harm must be 'serious' reflects the notion that not all forms of harm are

---

[26] See for example Rosemary Byrne, 'James C. Hathaway and Michelle Foster. The Law of Refugee Status' (2015) 26 EJIL 564; Hugo Storey, 'The Law of Refugee Status, 2nd edition: Paradigm Lost?' (2015) 27 IJRL 348; Audrey Macklin, 'The Law of Refugee Status 2d ed.' (2017) 39 Hum Rts Q 220.
[27] Hathaway and Foster, *Refugee Status* (n 16) 183.   [28] Ibid 205.
[29] Universal Declaration of Human Rights (adopted 10 December 1948) UNGA Res 217 A(III) (UDHR).
[30] International Covenant on Civil and Political Rights (adopted 16 December 1966, entered into force 23 March 1976) 999 UNTS 171 (ICCPR).
[31] International Covenant on Economic, Social and Cultural Rights (adopted 16 December 1966, entered into force 3 January 1976) 993 UNTS 3 (ICESCR).
[32] Convention on the Rights of the Child (adopted 20 November 1989, entered into force 2 September 1990) 1577 UNTS 3 (CRC).
[33] Erika Feller, 'Statement by Ms. Erika Feller, Director, Department of International Protection, on the Refugee Definition' (Brussels, Strategic Committee for Immigration, Frontiers and Asylum (SCIFA), 6 November 2002) 3 <http://www.refworld.org/docid/3dee02944.html> accessed 4 April 2019, partially quoted in Hugo Storey, 'What Constitutes Persecution? Towards a Working Definition' (2014) 26 IJRL 272, 277.
[34] *Islam v Secretary of State for the Home Department Immigration Appeal Tribunal and Another, ex parte Shah, R v* [1999] UKHL 20, [1999] AC 629 653 (Lord Hoffman).
[35] See further on the 'surrogacy paradigm' in Section 1.2.

sufficiently grave as to justify recognition of refugee status.[36] Importantly, serious harm can result from a single act, but may also result from an accumulation of measures, particularly in relation to denials of economic, social, and cultural rights.[37]

When the state itself sets out to cause serious harm, then it tends to follow that there is also a failure of state protection.[38] Where the claimant is exposed to serious harm as a consequence of the conduct of non-state actors of persecution, the question arises about the meaning of a failure of state protection. One approach has insisted that a person may have a well-founded fear of being exposed to serious harm at the hands of non-state actors on return but will be unable to establish a failure of state protection where the state has in place 'a system of domestic protection and machinery for the detection, prosecution and punishment' of relevant conduct and is 'able and willing to operate that machinery'.[39]

This approach, which has been criticised, suggests that

> an individual can be returned to his/her country of origin notwithstanding the fact that s/he has a well-founded fear of persecution on a Convention ground simply because the State did its best to operate the system of protection for the basic human rights of its nationals.[40]

This 'due diligence' approach has been widely rejected.[41] The better interpretation, which this book adopts, is that a failure of state protection will be established where the state is unable or unwilling to provide *effective* protection that would bring the fear of being persecuted below the well-founded standard,[42] not least because the 'central question in refugee law [is] whether an individual is in need of protection'.[43]

The preceding description of the being persecuted element is sufficient for the purposes of this chapter, but only begins to scratch the surface of the meaning of this central concept in international refugee law. Chapters 4–6 of this book examine the concept of being persecuted in greater depth, proposing a recalibrated human rights-based definition.

3.2.3 ... *for Reasons of*

Not all persons exposed to serious harm from which the state is unable or unwilling to provide effective protection are entitled to be recognised as refugees. Rather, only those who

---

[36] On the reasons why 'serious' harm is required in order to establish that a person is persecuted, see *DS (Iran)* [2016] NZIPT 800788, [114–27], in which the Tribunal draws on state practice, academic authority, as well as jurisprudence from leading common law jurisdictions.

[37] See Michelle Foster, *International Refugee Law and Socio-Economic Rights: Refuge from Deprivation* (CUP 2007) 93.

[38] For a detailed discussion of the kinds of circumstances that will reflect a failure of state protection, see Hathaway and Foster, *Refugee Status* (n 16) ch 4.

[39] *Horvath v Secretary of State for the Home Department* [2000] UKHL 37, [2001] AC 489 (Lord Clyde).

[40] Hélène Lambert, 'The Conceptualisation of "Persecution" by the House of Lords: *Horvath vs Secretary of State*' (2001) 13 IJRL 16, 28.

[41] See Goodwin-Gill and McAdam, *The Refugee in International Law* (n 20) 10–12; Hathaway and Foster, *Refugee Status* (n 16) 308–19 with extensive references to academic and judicial rejection of the approach.

[42] Hathaway and Foster, *Refugee Status* (n 16) 315–19 and express adoption of this approach in *Refugee Appeal No. 71427/99* RSAA (16 August 2000).

[43] Hathaway and Foster, *Refugee Status* (n 16) 310.

face being persecuted for reasons of their race, religion, nationality, membership of a particular social group, or political opinion will satisfy the refugee definition.

Two points require consideration. First, it is important to address the question of whether the motives of the actor of persecution are at all relevant to the assessment of whether a person's fear of being persecuted is for a Convention reason. Second, the standard of causation must be addressed, and here approaches such as those requiring that the Convention reason be 'the essential and significant reason' are contrasted with approaches acknowledging that multiple factors may contribute to the predicament of a person who is accurately described as being persecuted. Motivation and standard of causation are addressed in turn below.

3.2.3.1 Motivation

The question of the motivation of the actor of persecution is not settled in international refugee law, although leading doctrinal authorities as well as the UNHCR share the view that motivation is not a necessary condition, as will be set out below. The issue has featured in international refugee law for decades,[44] but divergence across jurisdictions continues to affect how refugee status is determined, including in the context of 'natural' disasters and climate change. Hence, although the contours of discussion have not evolved significantly in recent years, a brief treatment is in order.

The view that the motivation of the actor of persecution must be established finds expression in jurisprudence. *Elias Zacarias* is often identified as the leading US case advancing the motivation-based approach to understanding nexus. Here, the Supreme Court of the United States first recalled that:

> In construing statutes, 'we must, of course, start with the assumption that the legislative purpose is expressed by the ordinary meaning of the words used'.[45]

The Court then proceeds to note that:

> Elias-Zacarias objects that he cannot be expected to provide direct proof of his persecutors' motives. We do not require that. But *since the statute makes motive critical*, he must provide some evidence of it, direct or circumstantial.[46]

The majority opinion, delivered by Scalia J, does not offer any authority for why the words 'on account of', which replace 'for reasons of' in the domestic transposition of Article 1A(2) of the Refugee Convention, must be understood as making motive critical. The dissenting opinion of Justice Stevens disagrees with the assertion of the majority that the statute 'makes motive critical',[47] but does not engage in an alternative exercise of statutory interpretation, asserting instead that the conduct of the claimant *did* in fact amount to an expression of political opinion, and therefore the causal nexus was satisfied.

---

[44] See for example Karen Musalo, 'Irreconcilable Differences? Divorcing Refugee Protections from Human Rights Norms' (1993) 15 Mich J Intl L 1179; James Hathaway and Michelle Foster, 'The Causal Connection ("Nexus") to a Convention Ground' (2003) 15 IJRL 461.

[45] *Immigration and Naturalization Service v Elias Zacarias* [1992] 502 US 478 (USSC, Jan. 22, 1992) 482, quoting from *Richards v United States*, 369 US 1, 9 [1962].

[46] Ibid 483 (emphasis added).   [47] Ibid 489.

A similar view is found in Australian judicial opinion interpreting the refugee definition. The view of the Full Federal Court of Australia is widely cited:

> Persecution involves the infliction of harm, but it implies something more: an element of an attitude on the part of those who persecute which leads to the infliction of harm, or an element of motivation (however twisted) for the infliction of harm. People are persecuted for something perceived about them or attributed to them by their persecutors. Not every isolated act of harm to a person is an act of persecution.[48]

Again, this approach seeks to imagine the reasons driving actors of persecution to engage in certain acts of persecution, even if the focus here is on the meaning of 'persecution' itself, rather than the nexus clause. That locating the element of the refugee definition where 'motivation' is supposed to reside yields different results across jurisdictions in itself reflects a lack of clarity in the approach.

Doctrine developed by leading scholars of international refugee law rejects the notion that motive is a necessary condition for establishing nexus to a Convention reason. Goodwin-Gill and McAdam,[49] as well as Hathaway and Foster,[50] acknowledge that motive may be sufficient evidence to establish a nexus to a Convention ground, but it is not a necessary condition. Similarly, having regard to the object and purpose of the Convention 'to provide international protection to those in need',[51] Zimmermann and Mahler assert that:

> The determination of the causal connection between acts of persecution on the one hand, and one or more of the 1951 Convention grounds on the other, therefore needs to be made on the basis of an objective assessment of the underlying reasons for persecution, rather than being based on the subjective motivation of the respective persecutor.[52]

Hathaway and Foster identify a number of cases where the view that motive is central to persecution for a Convention reason has precluded recognition of refugee status.[53] Not only does this approach fail to recognise the wider social context within which a person's fear of being persecuted develops, but the authors also point to the plain evidentiary challenge of establishing motive in the context of RSD. The difficulty in establishing motive is highlighted by the High Court of Australia:

> [T]here is a special reason in the context of the Convention to refrain from importing concepts of personal motivation as essential to the context. By definition, the Convention will ordinarily be invoked in a foreign country where an inquiry into the motives and feelings of the alleged 'persecutors' will be extremely difficult or impossible to perform.[54]

---

[48] *Ram v Minister for Immigration and Ethnic Affairs*, [1995] 130 ALR 213 (Aus. FFC, Jun. 27, 1995) at 568, cited in Hathaway and Foster, *Refugee Status* (n 16) 369.
[49] Goodwin-Gill and McAdam, *The Refugee in International Law* (n 20) 101.
[50] Hathaway and Foster, *Refugee Status* (n 16) 368.
[51] Andreas Zimmermann and Claudia Mahler, 'Article 1A, para. 2 (Definition of the Term "Refugee"/ Définition du Terme "Réfugié")', in A Zimmermann (ed), *The 1951 Convention Relating to the Status of Refugees and its 1967 Protocol: A Commentary* (OUP 2011) 373.
[52] Ibid 374.   [53] See Hathaway and Foster, *Refugee Status* (n 16) 369–72.
[54] *Chen Shi Hai* (Aus. HC, 2000) at 313 [64], cited in Hathaway and Foster, *Refugee Status* (n 16).

In contrast to the view that motive is critical, senior courts in Australia,[55] New Zealand,[56] and the United Kingdom[57] have expressed more concern with the predicament of the claimant. The first manifestation of this approach builds on the understanding of 'persecution' as serious harm + the failure of state protection, presented above. Under this approach, the motive of the actor who inflicts harm may be evidence that a person's fear of being persecuted is for a Convention reason, but the kinds of circumstances in which a person may accurately be described as being persecuted for a Convention reason are far wider, and encompass at least situations where the state fails to protect an individual, and that failure can be linked to a Convention reason. Lord Hoffman's formulation in *Islam* is by now widely cited:

> [S]uppose that the Nazi government in those early days did not actively organise violence against Jews, but pursued a policy of not giving any protection to Jews subjected to violence by neighbours. A Jewish shopkeeper is attacked by a gang organised by an Aryan competitor who smash his shop, beat him up and threaten to do it again if he remains in business. The competitor and his gang are motivated by business rivalry and a desire to settle old personal scores, but they would not have done what they did unless they knew that the authorities would allow them to act with impunity.[58]

Thus, the direct actor of persecution may, as Hoffman observes, be motivated by purely personal and economic reasons, but is empowered to act because of the lack of state protection available to the claimants. Where the failure of state protection is connected to a Convention ground, it is not necessary to establish the motive of the actor who directly threatens or inflicts serious harm. In other words, the bifurcated approach does not say 'find motivation either within the serious harm limb or the failure of state protection limb', but rather that motivation is not determinative of nexus. The state may also lack a 'motivation' to cause harm, and may, as Hathaway expresses the approach,[59] 'not be bothered' to fulfil its obligations to protect a person or group of people because of who they are.

Emerging from this clear and principled approach is the interpretation of the nexus clause as enjoining decision-makers to look to the *predicament* of the claimant, in express contrast to focusing on the motivation of the actor of persecution.

Notwithstanding its now firmly established statutory interpretation that requires evidence that a Convention reason is 'the essential or significant reason' for a person's well-founded fear of being persecuted, Australia has also produced jurisprudence supporting a predicament approach, as Madgwick J explains:

> The question conventionally asked has been: Is the motivation of the persecutor the actual or perceived political opinion of the claimant? A more practical and properly inclusive question would appear to be: Is it the claimant's actual or perceived political opinion that accounts for the persecution the claimant fears? The latter question includes the former and is a closer paraphrase of the actual Convention language. It also better fastens attention on the necessity, in the interests of the vindication of human dignity, to rescue the claimant from the fearful *predicament* in which the combination of

---

[55] *NACM of 2002 v Minister for Immigration and Multicultural and Indigenous Affairs* [2003] FCA 1554, [63], cited in Foster, *International Refugee Law and Socio-Economic Rights* (n 37) 280 fn 193.
[56] *Refugee Appeal No. 72635/01 RSAA* (6 September 2002).   [57] *Islam* (n 34).   [58] Ibid
[59] James Hathaway, 'Food Deprivation: A Basis for Refugee Status?' (2014) 81 Sociol Res 327, 336 <http://repository.law.umich.edu/articles/1076/> accessed 4 April 2019.

his/her political opinion (or other Convention protected attribute) and the lack of effective state protection of the right to express such opinion puts him or her.[60]

Similarly, in *Refugee Appeal No. 72635/01*, the Refugee Status Appeals Authority of New Zealand expressly endorses the bifurcated approach:

> Looking first at the language of the Refugee Convention, the 'for reasons of' clause relates not to the word 'persecuted' but to the phrase 'being persecuted'. The employment of the passive voice ('being persecuted') establishes that the causal connection required is between a Convention ground and *the predicament of the refugee claimant*. The Convention defines refugee status not on the basis of a risk 'of persecution' but rather 'of being persecuted'. The language draws attention to the fact of exposure to harm, rather than to the act of inflicting harm. The focus is on the reasons for the claimant's predicament rather than on the mindset of the persecutor, a point forcefully recognised in Chen Shi Hai v Minister for Immigration and Multicultural Affairs (2000) 201 CLR 293 at [33] and [65] (HCA). At a practical level the state of mind of the persecutor may be beyond ascertainment even from the circumstantial evidence.[61]

The clear advantage of the predicament approach to nexus is that, building on a human rights-based interpretation of being persecuted, it explains why women, children, and other individuals fearing harm from non-state actors are appropriately recognised as falling within the refugee definition when they cross an international border and their fear of being persecuted on return is well-founded.

This approach, which is now regarded as the preferred approach by the UNHCR[62] and many legal doctrinal scholars,[63] relegates intention to the sidelines and focuses expressly on the holistic understanding of the circumstances from which the claimant has fled, and in which she will find herself on return. Hathaway and Foster frame the distinction thus:

> Framed simply, if a Convention ground explains why the applicant is exposed to the risk of being persecuted, is that sufficient to establish that there is a causal connection between a Convention ground and the reason for the applicant's well-founded fear of being persecuted?[64]

Their positive answer is that:

> This 'predicament approach' focuses attention not simply on the intent of the persecutor or of the state failing to protect, but more broadly on the reason for exposure to the risk. As the Federal Court of Australia concluded in the conscription context, 'even if a law is a law of general application, its impact on a person who possesses a Convention-related attribute can result in a real chance of persecution for a Convention reason'. This follows

---

[60] *NACM of 2002* (n 55) [63], cited in Foster, *International Refugee Law and Socio-Economic Rights* (n 37) 280 fn 193 (emphasis added).
[61] *Refugee Appeal No. 72635/01* (n 56) [168] (emphasis added).
[62] See for example UNHCR, 'Guidelines on International Protection No 12: Claims for Refugee Status related to Situations of Armed Conflict and Violence under Article 1A(2) of the 1951 Convention and/or 1967 Protocol Relating to the Status of Refugees and the Regional Refugee Definitions' (2 December 2016) HCR/GIP/16/12 [32] <http://www.refworld.org/docid/583595ff4.html> accessed 4 April 2019.
[63] See for example Hathaway and Foster, *Refugee Status* (n 16) 376–82, citing substantial jurisprudence from common law jurisdictions; Jason Pobjoy, *The Child in International Refugee Law* (CUP 2017) 160–64.
[64] Hathaway and Foster, *Refugee Status* (n 16) 376.

from the fact that 'the equal application of the law to all persons may impact differently on some of those persons' and the 'result of the different impact might be such as to amount to persecution for a Convention reason'.[65]

The focus on the predicament of the claimant set out above clearly integrates with a human-rights-based approach to the refugee definition. In particular, principles of non-discrimination under international human rights law do not regard intention as a necessary condition for the experience of discrimination,[66] and a predicament-based approach acknowledges both direct as well as indirect forms of discrimination as potentially establishing a nexus to one or more of the five Convention grounds.[67]

3.2.3.2 Standard of Causation

In addition to the question of motivation, the standard of causation has also vexed judicial authorities and legal doctrinal scholars seeking an accurate interpretation of the refugee definition.[68] There is both a narrow and a broader view of the standard of causation required to establish the nexus between being persecuted and one or more of the five Convention reasons.

In Australia, a narrow approach has been incorporated into the domestic legal framework. Importantly, and of relevance to the review of jurisprudence set out below, Australia now imposes a statutory definition of the nexus clause. Section 5J(4) of the Migration Act 1958 provides:

> (4) If a person fears persecution for one or more of the reasons mentioned in paragraph (1)(a):
>
> (a) that reason must be the *essential and significant reason*, or those reasons must be the essential and significant reasons, for the persecution; and
> (b) the persecution must involve serious harm to the person; and
> (c) the persecution must involve systematic and discriminatory conduct. [emphasis added]

Tests such as these, which require a very high degree of causation, have been largely rejected in other jurisdictions in favour of a 'contributory cause' approach, which recognises that a person may be exposed to serious harm for many reasons, including on account of her civil or political status. This approach is endorsed by the UNHCR[69] and is clearly articulated in the Michigan Guidelines:

> In view of the unique objects and purposes of refugee status determination, and taking account of the practical challenges of refugee status determination, the Convention ground need not be shown to be the sole, or even the dominant cause of the risk of being persecuted. It need only be a contributing factor to the risk of being persecuted.

---

[65] Ibid 378, citing a number of cases from the Federal Court of Australia.
[66] See for example Foster, *International Refugee Law and Socio-Economic Rights* (n 37) 283.
[67] See discussion in Hathaway and Foster, 'The Causal Connection' (n 44) 465.
[68] See for example the discussion in Zimmermann and Mahler, 'Definition of the Term "Refugee"' (n 51) 372–73, addressing 'but for', 'predominant cause', and 'contributing cause' approaches.
[69] See list of Guidelines on International Protection in which the 'contributory cause' approach is adopted in Hathaway and Foster, *Refugee Status* (n 16) 389 fn 160.

If, however, the Convention ground is remote to the point of irrelevance, refugee status need not be recognized.[70]

The New Zealand Refugee Appeals Authority cogently articulated this approach in *Refugee Appeal No. 72635/01*:

> Multiple causes and evidentiary gaps, so characteristic of refugee law, pose serious challenges to the successful determination of causation. Ultimately, causation standards in the refugee context must operate clearly and consistently to accommodate both multiple causes and evidentiary insufficiency, if not ambiguity.
>
> We are of the view that it is sufficient for the refugee claimant to establish that the Convention ground is a contributing cause to the risk of 'being persecuted'. It is not necessary for that cause to be the sole cause, main cause, direct cause, indirect cause or 'but for' cause. It is enough that a Convention ground can be identified as being relevant to the cause of the risk of being persecuted. However, if the Convention ground is remote to the point of irrelevance, causation has not been established.[71]

Similarly, the US Court of Appeals for the Seventh Circuit was able to look beyond the apparent 'personal motivation' in the case of a threatened 'honour killing' to find that the claimant's gender, and therefore membership of a particular social group in a society that supported such conduct, to be a reason for her well-founded fear of being persecuted:

> The man who does the killing may have a personal motivation in the sense that he is angry that his sister has dishonored the family, or he may regret the need to take such an irrevocable step. Either way, he is killing her because society has deemed that this is a permissible – maybe in some eyes the only – correct course of action and the government has withdrawn its protection from the victims.[72]

Citing a wide range of authorities, Hathaway and Foster conclude that:

> In light of the myriad practical challenges in assigning particular weight to a Convention ground, and the lack of any principled or ethical basis for denying protection so long as a Convention ground is a contributing factor in establishing risk, it is logical that the overwhelming judicial preference is to adopt a straightforward 'one factor' test.[73]

This book expressly adopts a predicament approach as the most persuasive interpretation of the relationship between being persecuted and the Convention grounds in international refugee law doctrine. It accepts the argument that a contributory cause standard accurately reflects the degree of connection between the Convention reasons and a person's well-founded fear of being persecuted.

---

[70] Michigan Guidelines on Nexus to a Convention Ground (2002) 23 Mich J Intl L 211, cited in Hathaway and Foster, 'The Causal Connection' (n 44) 476. The Michigan Guidelines are not a source of law, but rather reflect the consensus of a group of experts in international refugee law, who convene to discuss a specific theme in depth with a view to developing guidelines. For more on the Guidelines, see <http://www.law.umich.edu/centersandprograms/refugeeandasylumlaw/Pages/colloquiumandmichguidelines.aspx> accessed 4 April 2019.
[71] *Refugee Appeal No. 72635/01* (n 56).
[72] *Sarhan* (USCA, 7th Cir., 2010) at 608, cited in Hathaway and Foster, *Refugee Status* (n 16) 385.
[73] Hathaway and Foster, Refugee Status (n 16) 389.

3.2.4 ... Race, Religion, Nationality, Membership of a Particular Social Group, or Political Opinion

Although there is extensive doctrine and jurisprudence concerning the meaning of each of these Convention grounds, the determination of a person's civil or political status in no way differs according to the circumstances of the case. Hence, the methodology for determining whether a person is a member of a particular social group, holds a certain political opinion or religious belief, or is identifiable as possessing a distinct nationality will not differ according to whether the person finds herself in a dictatorship, armed conflict, disaster situation, or any combination of the three. Hence, the book does not engage with this element of RSD.

3.2.5 Internal Relocation

Finally, a person will not be eligible for recognition of refugee status if she only has a well-founded fear of being persecuted in one part of the country, provided it can be established that relocation to another part of the country is reasonable in all the circumstances.[74] Although, as with most other elements of the refugee definition, there are differing interpretations of the scope of the internal relocation element,[75] the fact remains that if safe, legal means of relocating internally are available to the claimant, and the conditions awaiting the claimant in that part of the country do not entail a risk of a serious denial of human rights, then the person will not be eligible for recognition of refugee status.[76] The underlying rationale of the internal relocation alternative is the doctrine of surrogate protection, which holds that the purpose of the Convention is to provide international protection in those cases where there is a failure of protection in the claimant's country of origin.[77] Although there are differences in approach, with some cases incorporating the question of internal relocation within the assessment of whether a person has a well-founded fear of being persecuted for a Convention reason,[78] the more coherent approach endorsed by the UNHCR[79] and as articulated by Hathaway and Foster is to consider the question of internal relocation only after a person has established a prima facie case for

---

[74] UNHCR, 'Guidelines on International Protection: "Internal Flight or Relocation Alternative" within the Context of Article 1A(2) of the 1951 Convention and/or 1967 Protocol Relating to the Status of Refugees' (23 July 2003) HCR/GIP/03/04, [7].

[75] For detailed treatment of this element of RSD, see Jessica Schultz, *The Internal Protection Alternative in Refugee Law: Treaty Basis and Scope of Application under the 1951 Convention Relating to the Status of Refugees and Its 1967 Protocol* (Brill 2018) and Bríd Ní Ghráinne, 'The Internal Protection Alternative Inquiry and Human Rights Considerations – Irrelevant or Indispensable?' (2015) 27 IJRL 29.

[76] This paraphrases the UNHCR Guidelines on internal relocation (n 74).

[77] James Hathaway and Michelle Foster, 'Internal Protection/Relocation/Flight Alternative as an Aspect of Refugee Status Determination' in Erika Feller, Volker Türk and Frances Nicholson (eds), *Refugee Protection in International Law: UNHCR's Global Consultations on International Protection* (CUP 2003) 359.

[78] *Januzi v Secretary of State for the Home Department* [2006] UKHL 5, [2006] 2 AC 426 [8], cited in Ní Ghráinne, 'Irrelevant or Indispensable?' (n 75) 32.

[79] UNHCR, '"Internal Flight or Relocation Alternative"' (n 74) [7].

recognition as a refugee,[80] with reference to the 'unwilling to avail himself of the protection of [the home] country' element of Article 1A(2).[81]

3.3 Structure of the Review

Each of these elements of the refugee definition, as understood according to prevailing interpretations, has been identified as presenting challenges to the recognition of refugee status of persons displaced across borders in the context of 'natural' disasters and climate change. It is to this body of jurisprudence that the chapter now turns. Cases are presented according to the following taxonomy. First, cases where disasters and climate change are peripheral to the claim are briefly identified. Next, 'Category 1' cases, which frame a fear of being persecuted on return as a fear of 'indiscriminate adversity due to the forces of nature' are considered. Then, 'Category 2' cases, which reflect scenarios where identifiable actors of persecution may be expected to directly and intentionally inflict serious harm in the context of disasters and climate change are briefly summarised. Finally, 'Category 3' cases, which relate to other *ex ante* and *ex post* failures of state protection are examined closely, as these cases present the clearest doctrinal challenges relating to refugee status determination in the context of disasters and climate change. The Appendix consolidates the findings of the review in a taxonomy of the kinds of circumstances in which a person may establish a well-founded fear of being persecuted for a Convention reason in the context of 'natural' disasters and climate change. This table also contains references to legal doctrinal scholarship highlighting potential circumstances, even when they have not been established by jurisprudence.

3.4 Disasters and Climate Change Are Peripheral to the Majority of Claims in the Review

Most cases in this category contain references to the occurrence of natural hazard events as 'relevant background' within judicial decisions. A smaller number of cases contain references to disasters that affected the claimant in a particular way, but which is neither relied upon by the claimant, nor addressed in the determination of eligibility for recognition of refugee status. Cases falling into each of these two subcategories are described briefly below.

The vast majority of cases that arise in the context of 'natural' disasters and climate change do not reflect an express concern articulated by the claimant about being exposed to disaster-related harm if returned. Indeed, in most of the cases surveyed, the claimant does not adduce any evidence pertaining to disasters and climate change at all. Rather, environmental degradation or a particular hazard event is referenced by the judicial authority as 'relevant background'. More often than not, environmental degradation and disasters are not even identified by the judicial authority as being expressly relevant, but rather reference appears with quotations from country of origin reports addressing country conditions more generally. Thus, although references to environmental degradation and

---

[80] Hathaway and Foster, 'Internal Protection/Relocation/Flight Alternative' (n 77) 370.
[81] A point Hathaway and Foster make clearly in *Refugee Status* (n 16) 332.

disasters feature in cases relating to Somalia,[82] Afghanistan,[83] Bangladesh,[84] China,[85] Iraq,[86] Sudan,[87] Sri Lanka,[88] India,[89] East Timor,[90] Indonesia,[91] Honduras,[92] Kenya,[93] and

---

[82] *MA (Somalia) v Secretary of State for the Home Department* [2009] EWCA Civ 4 [appellant's mother disappeared in the 2004 Indian Ocean tsunami]; *HH and Others (Mogadishu- Armed Conflict- Risk) Somalia v Secretary of State for the Home Department CG* [2008] UKAIT 00022 [multiple references to disaster, drought, famine, food aid, flood]; Borgarting Court of Appeal Decision of 23 September 2011– *Abid Hassan Jama v Utlendingsnemnda*, 10-142363ASD-BORG/01, Norway [drought a significant factor precluding internal relocation].

[83] *GS (Article 15(c)- Indiscriminate Violence) Afghanistan v Secretary of State for the Home Department CG* [2009] UKAIT 00044 [reference to disasters and poor food production have greater impact on mortality than conflict in Afghanistan]; *HK and Others Afghanistan v Secretary of State for the Home Department CG* [2010] UKUT 378 [reference to drought contributing to food insecurity forcing families to send children into the streets to beg for food and money, in case about the return of children to Afghanistan]; *AK (Article 15(c)) Afghanistan v Secretary of State for the Home Department CG* [2012] UKUT [general reference to 'floods and natural disasters' as factors contributing to difficulties faced by IDPs]; *14 (Kabul - Pashtun) Afghanistan v Secretary of State for the Home Department CG* [2002] UKIAT 05345 [difficulty accessing water in Kabul as a consequence of four years of drought]; *Refugee Appeal No. 76191* RSAA (12 August 2008) [on the interaction between drought and conflict].

[84] *Refugee Appeal No. 76128* RSAA (26 March 2008) [reference to environmental degradation as a factor fuelling conflict].

[85] *RRT Case No. 071576767* [2007] RRTA 223 (25 September 2007); *RRT Case No. 060819216* [2006] RRTA 215 (22 December 2006); *RRT Case No. 1001270* [2010] RRTA 289 (20 April 2010); *RRT Case No. 0903266* [2009] RRTA 850 (18 September 2009); *Refugee Appeal No. 3/91 Re ZWD* RSAA (20 October 1992) [passing references to environmental degradation in country of origin information].

[86] *HM and Others (Article 15(c)) Iraq CG v the Secretary of State for the Home Department*, [2012] UKUT 00409 [on drought affecting conditions for internal relocation to the KRG in Iraq]; *Refugee Appeal No. 76457* RSAA (15 March 2010) [lack of water due to drought for IDPs in Dahuk].

[87] *Refugee Appeal No. 76074* RSAA (22 November 2007) [the relationship between drought, desertification, and conflict]; *LM (Relocation - Khartoum - AE Reaffirmed) Sudan v Secretary of State for the Home Department* [2005] UKAIT 00114 [reference to claimant's ethnic group, the Beja, having suffered long-term discrimination making them vulnerable to malnutrition, famine, and contagious disease in Eastern Sudan]; *Refugee Appeal No. 75655* RSAA (29 September 2006) [famine, drought, and relocation to urban slums].

[88] *RRT Case No. 1203764* [2012] RRTA 312 (18 May 2012) [judge enquiring whether claimant had been affected by the 2004 Indian Ocean tsunami – claimant recognised as a refugee on entirely unrelated grounds].

[89] *RRT Case No. 0808164* [2009] RRTA 519 (4 June 2009) [passing reference to environmental pollution and degradation in relation to conflict between religious groups].

[90] *RRT Case No. 071204406* [2007] RRTA 86 (27 April 2007) [references to drought, tsunami, food insecurity in a conflict situation].

[91] *RRT Case No. 060793741* [2007] RRTA 3 (11 January 2007); *RRT Case No. 061051351* [2007] RRTA 15 (2 February 2007) [in both cases, public dissatisfaction with the pace of reconstruction following the 2004 Indian Ocean tsunami, accompanied by recognition of the role played by the disaster in ending the conflict].

[92] *Chacon v INS*, 341 F.3d 533, United States Court of Appeals for the Sixth Circuit, 18 August 2003 [identification by expert witness of the role played by 1998 Hurricane Mitch in the deterioration of public security in Honduras: 'The professor explained that the economic consequences of this natural disaster led to a general rise in the level of violence in Honduras, and subsequently a steep increase in the number of extra-judicial murders committed by Honduran security forces and/or paramilitary groups, specifically targeting young men with tattoos, who were assumed to be gang members involved in criminal activities'].

[93] *RRT Case No. 0903027* [2009] RRTA 1102 (28 October 2009) [drought referred to in country of origin information about the causes of inter-ethnic conflict in case about a person of Luo ethnicity fearing persecution by members of the Kikuyu ethnic group].

Zimbabwe,[94] there tends to be little or no judicial consideration of the relevance of such events or processes to the individual claim.

The existence of a substantial number of such cases where disasters feature within the relevant background to a claim for recognition of refugee status may repay more detailed consideration, for example in further research relating to the intersection between climate change, disasters, and conflict.[95] However, this body of jurisprudence does not assist in determining the kinds of circumstances in which an individual may establish a well-founded fear of being persecuted for a Convention reason in the context of 'natural' disasters and climate change.

The review also revealed a second subcategory of 'peripheral' cases where the claimant's evidence reflected the fact that she had been adversely affected in a disaster before travelling to the host county, even though the event was not relied upon in support of the claim itself. In some cases, the event formed part of a much longer personal narrative, and in some cases had occurred several years before the claimant's departure from her home country.[96] In other cases, the claimant had left her home country within months of the event unfolding, and in some cases identified it as a relevant factor in the decision to leave.[97] Although these cases contribute further insight into the range of individual displacement scenarios that are

---

[94] See references in Section 3.6.3 on discriminatory distribution of food relief in the context of drought and economic collapse.

[95] Relevant literature and UN statements on this intersection includes: Peter Gleick, 'Global Climatic Change and International Security' (1990) 1 Col J Int'l Envtl L & Pol'y 41; Peter Gleick, 'Water, Drought, Climate Change, and Conflict in Syria' (2014) 6 Weather Clim Soc 331; Jon Barnett 'Security and Climate Change' (2003) 13 Global Environ Chang 7; UN Security Council, '5663rd meeting' (17 April 2007) S/PV.5663; UN Security Council, '6587th meeting, (Resumption 1)' (20 July 2011) S/PV.6587; UN General Assembly, 'Climate Change and its Possible Security Implications: Report of the Secretary-General' (11 September 2009) A/64/350. See also Sanjula Weerasinghe, 'In Harm's Way: International Protection in the Context of Nexus Dynamics between Conflict or Violence and Disaster or Climate Change' (2018) UNHCR Legal and Protection Policy Research Series PPLA/2018/05 <https://www.unhcr.org/5c1ba88d4.pdf> accessed 16 April 2019.

[96] *R v Secretary of State for the Home Department, ex parte Ahmed Aissaoui* United Kingdom High Court (England and Wales) (11 November 1996) CO/1391/96 [family killed in earthquake in Algeria and he had nobody left. Claimed asylum in the United Kingdom an indeterminate time later, fearing the FIS group]; *RRT Case No. 1211848* [2012] RRTA 933 (16 October 2012) [claimant's identity documents destroyed in a cyclone in Vietnam in 2005. Authorities refused to renew, telling him to return to the North. Without identity documents, fell into a downward spiral of criminality and gang affiliation. Claimed asylum in Australia in 2011 on the basis of difficulties associated with not having identity documents].

[97] *TN and SN v Denmark* [2011] ECHR 94 [claimants entered Denmark in September 2005 after their fishing boat had been destroyed in the December 2004 Indian Ocean tsunami – basis of claim related to fear of Liberation Tigers of Tamil Eelam (LTTE)]; *RRT Case No. 1215708* [2012] RRTA 1084 (6 December 2012) [Afghan unable to farm due to drought, consequently forced to sell chickens, requiring him to travel on a road controlled by the Taliban]; *RRT Case No. 1209786* [2012] RRTA 1058 (29 November 2012) [claimant came to Australia in February 2006 after 2005 earthquake in Pakistan. Visa expired in May 2006. Claimant apprehended in 2012 and detained. Claimed asylum. Claimant references wanting to help family by sending money to them as they had property damage in the earthquake. Basis of claim was fear of retribution for refusing to marry someone]; *RRT Case No. 1203936* [2012] RRTA 1070 (14 December 2012) [earthquake in China in 2008 caused damage to claimant's restaurant, leading to its closure. He also claimed to have lost his identity documents. Entered Australia as a visitor in February 2011 and approximately six months after his visitor visa expired, claimed asylum citing a fear of being persecuted on the basis of his membership in Falun Gong]; *RRT Case No. 1002915* [2011] RRTA 615 (14 July 2011).

more or less connected to 'natural' disasters and climate change, they are not examined in detail here as they do not provide insight into the kinds of circumstances in which an individual may establish a well-founded fear of being persecuted for a Convention reason in the context of 'natural' disasters and climate change.

3.5 Category 1: Indiscriminate Adversity due to the Forces of Nature

Cases falling within this category reflect the powerful hold that the hazard paradigm, described in Chapter 2, continues to exercise over judicial understanding of disasters and climate change. In this category, adversity that a person has been exposed to or fears being exposed to upon return has nothing to do with the broader social context, but rather reflects misfortune due to the indiscriminate forces of nature.

The case of *Mohammed Motahir Ali v Minister of Immigration, Local Government and Ethnic Affairs*[98] is identified by McAdam in support of her contention that the Refugee Convention generally does not apply in the context of disasters.[99] The facts of the case are not set out in detail, as the focus of the Federal Court was a technical question of procedure under the Migration Act 1958. However, in recounting the history of the claim, Whitlam J records that the claimant entered Australia on a visa in July 1988 and in December 1989 lodged documents including a form entitled 'Application for Refugee Status in Australia'. On 6 February 1990, the claimant's solicitors submitted a nine-page statement in which the claimant asked for his claim to be considered as an 'economic refugee'. One month later, the claim was refused as manifestly unfounded. Several subsequent applications and negative decisions ensued, culminating in a refusal to recognise refugee status in January 1991. In responding to this outcome, the claimant's solicitors sought to rely upon a policy set out in Ministerial Press Statement (MPS) 15/91, which included provision for a humanitarian category. Purporting to appeal against the refusal, the claimant's solicitors, without having been instructed by the claimant, referred in a letter of May 1991 to 'the devastating impact of the floods and cyclones which have totally destroyed large parts of Bangladesh'. The claimant's solicitors conceded that the claimant 'cannot meet the technical definition according to DORS'. In subsequent representations, this time on the instruction of the claimant, the solicitors provided some very limited personal information about the claimant's fear of returning to Bangladesh in the aftermath of the cyclone:

> It is clear that the applicant would suffer intense personal hardship if he were forced to return to Bangladesh. It is submitted that the circumstances which the applicant has outlined of his family's conditions in their home town, which was devastated by the floods earlier this year, circumstances which the applicant would himself be subject to on his return, clearly would put him in a position of extreme and intense personal hardship. Under the circumstances I would therefore maintain that if it is not possible to sympathetically reappraise the compelling nature of the originally presented grounds, the applicant clearly can base an alternative case successfully on the recommended guidelines for humanitarian assessment. Please therefore sympathetically consider this application.

---

[98] *Mohammed Motahir Ali v Minister of Immigration, Local Government and Ethnic Affairs* [1994] FCA 887.
[99] Jane McAdam, *Climate Change, Forced Migration and International Law* (OUP 2012), 47 fn 44.

The claim articulated here is not based on the Refugee Convention, but on the humanitarian policy set out in MPS 15/91. Nonetheless, the Federal Court records a June 1992 decision of the Refugee Status Review Committee, including the conclusion that:

> The reasons for his departure from Bangladesh were of a purely economic nature and he came to Australia because of the better opportunities that this country afforded him. There is no doubt that the conditions caused by natural disasters in Bangladesh *impose hardship on all the residents of that country*. However, such circumstances, difficult though they may be, do not provide the basis for real fear of persecution for matters identified by the Convention. [emphasis added]

As a generalisation reflecting assumptions about 'natural disasters' that mirror those articulated at in the dominant view described in Chapter 1, the view of the Refugee Status Review Committee dismissing the possibility of refugee status arising in the context of disasters and climate change reflects a lack of anxious scrutiny to which each applicant for recognition of refugee status is entitled.[100] Having not considered the particular facts of the case, the replication of the paradigm-level rejection of disaster-related claims for recognition of refugee status amounts to little more than a reflexive reiteration of hazard-dominated thinking about the nature of disasters.

Another scenario reflecting adversity arising from the forces of nature comes from *RRT Case No. 0901657* (the 'Aceh tsunami case').[101] The case concerned an ethnically Chinese man of Christian faith who had lived in Jakarta, Indonesia, from 1977 to 1998 but who had relocated to Aceh following riots in Jakarta in 1998. He stayed in Aceh until shortly after the December 2004 tsunami, after which time he moved to Medan, Indonesia, before travelling to Australia in October 2008. He expressly identified the impact of the tsunami as his reason for seeking recognition of refugee status:

> Since June 1977 I live in Jakarta. I left Jakarta after Riot May 1998 and moved to Aceh. On December 2004 Tsunami came destroyed everything belong to me so I had to left that's place and moved to Medan.
>
> I was scare, sad and trauma because what kind of country, we work hard pay taxes but no one protect our rights not even tried it. That's why I moved to Aceh but bad luck Tsunami happened. I lost everything I came to Australia hoping find my safety and freedom. (I will send more details later on).[102]

The first-instance decision-maker had refused to grant a protection visa to the claimant, finding no evidence that the claimant had been personally targeted during the riots in 1998, and finding that, in the Tribunal's words, 'the applicant's loss of his possessions was a

---

[100] This notion of 'anxious scrutiny' in the context of international refugee law was first articulated by Lord Bridge of Harwich in *R v Secretary of State for the Home Department, Ex p Bugdaycay* [1987] AC 514 at 531F, where it is written: 'The most fundamental of all human rights is the individual's right to life and when an administrative decision under challenge is said to be one which may put the applicant's life at risk, the basis of the decision must surely call for the most anxious scrutiny.' The notion of 'searching scrutiny' appears to have been coined by Keane J in *S v Chief Executive of the Department of Labour and Anor* [2005] NZHC 439, HC AK CIV 2005-404-003360, drawing upon a series of cases commencing with *K v Refugee Status Appeals Authority* (HC Auckland, M 1586-SW99, February 2000) in which Anderson J posited at [40] that 'where the consequences of a wrong decision could be persecution of a most grave and inhumane nature, a reviewing court should look at an impugned decision with great care'.

[101] *RRT Case No. 0901657* [2009] RRTA 526 (11 June 2009).   [102] Ibid [21].

personal misfortune unrelated to a Convention ground'. Further, 'there was nothing in the evidence before her to suggest that the Indonesian authorities would be unwilling or unable to protect the applicant for a Convention reason'.[103]

Regrettably, the claimant did not attend the Tribunal on the date set for the hearing, and was therefore unable to provide additional evidence in support of his claim. Had he done so, he may have been able to answer a range of questions about his predicament and how it reflected a well-founded fear of being persecuted for a Convention reason on return to Indonesia. Such questions may have included substantially more robust details of his life as a Chinese Christian living on the majority Muslim island of Aceh. He may have explained where he lived and the kind of accommodation he lived in. He could have given evidence about his livelihood and whether that was affected by the tsunami. He could have given evidence about disaster response and recovery, including the treatment he received at the hands of both state and non-state actors. He could have explained why he moved to Medan and what his life was like there. Even if it had been established that he faced being persecuted if returned to Aceh, had he found a viable internal relocation alternative in Medan? Instead, on the basis of the very limited evidence provided, the Tribunal concluded:

> Similarly, whilst the Tribunal has had regard to the applicant's claim that he lost everything during the December 2004 Asian tsunami, the Tribunal observes that the applicant has not identified who the perpetrators of the alleged persecution in this instance were, or what was the selective or discriminatory conduct that gave rise to the applicant's fear. Nor has he identified how the essential and significant reason for the harm the applicant fears is related to any of the Convention grounds of race, religion, nationality, membership of a particular social group or political opinion. As a result, the Tribunal is not satisfied that the applicant had to flee Indonesia for his own safety.[104]

The hazard paradigm is again evident in this determination, as 'the December 2004 Asian tsunami' is framed purely as a force of nature. Of course, an oral hearing would have provided an opportunity for the claimant to highlight any wider social context that may have been relevant to the assessment of his claim.

The earliest case to consider the potential application of the Refugee Convention in the context of expressly climate change-related harm was *RRT Case No. N96/10806*.[105] The unrepresented claimant, a 35-year-old woman from Tuvalu, applied for a protection visa in Australia. She stated in her application:

> I left Tuvalu because I feared for my future as I found unsatisfactory means of supporting myself there. In Tuvalu there is only poverty for me.

In dismissing the claim, the Tribunal, expressly following Hathaway,[106] recognised the potential for socio-economic forms of harm to engage host state protection obligations under the Refugee Convention, provided that the causal nexus could be established between the harm feared and one of the five Convention reasons. In surveying the country of origin information, the Tribunal reviewed evidence suggesting that climate change was

---

[103] Ibid [25].   [104] Ibid [38].
[105] *RRT Case No. N96/10806* [1996] RRTA 3195 (7 November 1996). On the same day, in *RRT Case No. N95/09386* [1996] RRTA 3191 (7 November 1996), the Tribunal reached the same conclusion on similar facts relating to an unrepresented male applicant from Tuvalu who was married to an Australian citizen and had a child living in Australia.
[106] Hathaway, *Refugee Status* (n 25).

contributing to rising sea levels and more frequent and intense cyclones around Tuvalu. These manifestations of climate change were having a generalised adverse impact.

The Tribunal concluded that the claimant was not entitled to refugee status because any adverse socio-economic impacts she would face on return were not the result of persecution. Again relying on Hathaway, the Tribunal concluded that the environmentally related harm to which the claimant may be exposed upon return did not engage Australia's protection obligations because 'the environmental problem of the rise in the sea level around Tuvalu is not Convention related'. As the determination focuses exclusively on the natural hazards, it fails to enquire into the particular context in which such risks are engendered and may accrue. It is a clear example of how the hazard paradigm influences RSD, even if the outcome in this particular case may not have been altered by the more searching enquiry into the social context within which exposure and vulnerability to climate-change-related harm manifests.

A similar approach can be found in the case of *RRT Case No. 0907346*.[107] The Tribunal summarised the basis of the applicant's claim:

> The fear of what may happen if he returned to Kiribati was identified as not being able to work, support his family and eventually not having anywhere to go. Drinking water was polluted and incursion of sea water could remain for substantial periods, affecting fruit trees and other food crops. Fortnightly king tides flooded the area of the applicant's village and eventually the island would sink under water.[108]

The submissions of the applicant's legal representative are reproduced in the text of the determination.[109] The following legal arguments are advanced:

- Under existing laws, people affected by climate change are not recognised as a cognisable group in need of protection. However, existing protection visa laws can, and should, be creatively interpreted to accommodate climate change refugees.
- Climate change should be seen as a form of persecution which involves serious harm. The domestic legal definition of 'serious harm' is capable of being satisfied in light of the impact of rising sea levels, salination, and floods associated with climate change, which will cause the people affected significant economic hardship and threaten the applicant's capacity to subsist.
- The government of Kiribati is unable to protect the applicant from persecution, with reference to the standard set out by McHugh J in *Chan*.
- The people of Kiribati, and especially the claimant and others from parts of the island that are particularly affected by rising sea levels and salination, constitute a cognisable particular social group.
- The applicant's fear is well-founded owing to the large body of evidence establishing the reality of climate change and the vulnerability of low-lying Pacific island states like Kiribati to sea level change, violent storms, and ultimately submergence.

The determination provides details of the issues discussed at the oral hearing, including the concerns of the Tribunal regarding the lack of clear evidence of an actor of persecution driven by a motivation to cause harm for one of the five Convention reasons, in particular the lack of any apparent motivation on the part of high greenhouse-gas-emitting countries

---

[107] *RRT Case No. 0907346* [2009] RRTA 1168 (10 December 2009).   [108] Ibid [21].   [109] Ibid [22].

to harm the people of Kiribati.[110] The Tribunal records discussing evidence that the islands of Kiribati may become submerged within ninety years, and the applicant and his legal representative arguing that such anticipated submergence may well come much sooner.[111] Subsequent to the oral hearing, the applicant's legal representative made additional submissions that 'Australia's continued production of high levels of such pollution, in complete disregard for people on low lying islands, constitutes the relevant motivation to characterise climate change as persecution'.[112]

The appeal was dismissed. The Tribunal, following *Ram v MIEA and Anor*[113] and *Applicant A v MIEA*,[114] considered that the claimant had been unable to establish the necessary element of discriminatory motivation required for any harm feared to amount to persecution.

> In this case, the Tribunal does not believe that the element of an attitude or motivation can be identified, such that the conduct feared can be properly considered persecution for reasons of a Convention characteristic as required. It has been submitted that the continued production of carbon emissions from Australia, or indeed other high emitting countries, in the face of evidence of the harm that it brings about, is sufficient to meet this requirement. In the Tribunal's view, however, this is not the case. There is simply no basis for concluding that countries which can be said to have been historically high emitters of carbon dioxide or other greenhouse gases, have any element of motivation to have any impact on residents of low lying countries such as Kiribati, either for their race, religion, nationality, membership of any particular social group or political opinion. Those who continue to contribute to global warming may be accused of having an indifference to the plight of those affected by it once the consequences of their actions are known, but this does not overcome the problem that there exists no evidence that any harms which flow are motivated by one or more of the Convention grounds.[115]

The difficulties establishing a well-founded fear of being persecuted in the context of the adverse impacts of climate change on small island developing states in the South Pacific are made very clear by this case, which highlights difficulties in relation to the notion of being persecuted, the nexus clause, and the assessment of risk on return. Clearly, the more restrictive Australian approach that requires discriminatory motivation in order to establish a well-founded fear of being persecuted for a Convention reason makes the prospect of success for claimants seeking recognition of refugee status in this connection appear quite remote.

The argument advanced by the claimant reflects again the hazard paradigm as an epistemological foundation for RSD in this connection. Aware of the need to identify an actor of persecution, the claimant does not look to the social context, but rather seeks to anthropomorphise climate change by reference to the responsibility of rich countries for the bulk of greenhouse gas emissions.

These cases, along with a number of similar cases determined by Australian judicial authorities,[116] were influenced by the particular statutory interpretation of the refugee definition set out at Section 91 of the Migration Act 1998.[117] It has been noted at several

---

[110] Ibid [30], [37].   [111] Ibid [34], [37–40].   [112] Ibid [45].   [113] *Ram* (n 48).
[114] *Applicant A* (n 10).   [115] *RRT Case No. 0907346* (n 107) [51].
[116] *RRT Case No. 1004726* [2010] RRTA 845 (30 September 2010) [Tonga], *RRT Case No. N99/30231* [2000] RRTA 17 (10 January 2000) [Tuvalu].
[117] Now s 5J(4) of the Migration Act 1958.

points in the analysis of these determinations that an approach based more closely on international refugee law as reflected in doctrine and authoritative guidance produced by the UNHCR may not have resulted in a different outcome, but would have invited closer scrutiny of the claims. However, as the early determinations of the New Zealand Refugee Status Appeals Authority reflect, even when an expressly human-rights-based approach to refugee status determination is adopted, the hazard paradigm can easily misdirect attention away from the social context and towards the indiscriminate forces of nature.

The first climate-change-related claim to be considered in New Zealand was *Refugee Appeal No. 72186/2000*.[118] Summarising the basis of claim at paragraph 9, the RSAA noted:

> [T]he appellant complained, inter alia, that her family's house in Tuvalu is five feet from the sea and when the tide was high, the land would be submerged in water. The coastlands of the island were suffering erosion. Her husband had problems with his legs and could not work and her own legs were also 'not good'. Her biggest fear is that they would find it difficult to move around in the rising tide. There was also a shortage of drinkable water and medicine.

The claimant argued that she was eligible for refugee status. She identified the authorities in Tuvalu as actors of persecution, and argued that the state had failed to protect her, and that such failure was for reasons of race, nationality, and/or membership of a particular social group. The determination records the claim at paragraph 12:

> that the Tuvalu government failed in its duty of protecting the civil political, social, cultural and economic rights of the appellant for reasons of race (Tuvaluan), as nationals of Tuvalu (citizens) and as a member of a particular social group (defined as having no means to sustain themselves and survive).

The Authority dismissed the appeal, concluding at paragraph 16 that there was no causal nexus between the harm feared and one of the five Convention reasons:

> Clearly, none of the fears articulated by the appellant vis-à-vis her return to Tuvalu, can be said to be for reason of any one of the five Convention grounds ... This is not a case where the appellant can be said to be differentially at risk of harm amounting to persecution due to any one of these five grounds. All Tuvalu citizens face the same environmental problems and economic difficulties living in Tuvalu. Rather, the appellant is an unfortunate victim, like all other Tuvaluan citizens, of the forces of nature leading to the erosion of coastland and the family home being partially submerged at high tide.

In concluding that the claimant did not have a well-founded fear of being persecuted for a Convention reason, the Authority did not expressly state the definition of being persecuted, the accepted interpretation of the nexus clause, or its understanding of how a risk assessment is to be conducted in RSD claims. It had not been presented with a wide range of country of origin information relating to the socio-economic conditions in Tuvalu, and it did not examine in depth the impact of climate change and the occurrence of natural hazard events on people living there or the state's ability and willingness to protect the population from the associated risks. The Authority's implicit adoption of the hazard paradigm is clear. Seven days later, employing similar reasoning, the Authority dismissed the combined appeal of seven family members from Tuvalu.[119]

---

[118] *Refugee Appeal No. 72186/2000* RSAA (10 August 2000).
[119] *Refugee Appeal Nos. 72189–72195/2000* (17 August 2000).

Although this earlier New Zealand jurisprudence followed a similar course to the Australian cases outlined above, more recent determinations of the New Zealand Immigration and Protection Tribunal (NZIPT) demonstrate a substantial doctrinal and contextual depth, reflecting an intention on the part of the NZIPT to articulate a clear and principled methodology for determining refugee status in the context of 'natural' disasters and climate change. These cases are considered later in Section 3.7.

3.6 Category 2: Direct and Intentional Infliction of Harm

Cases falling within this category concern scenarios where individuals are exposed to acts of physical violence at the hands of the state and/or non-state actors. These cases are those most often cited by legal academics writing about the potential application of the Refugee Convention in the context of 'natural' disasters and climate change.

Three kinds of circumstances fall within this category of cases:

- The state intentionally causing environmental damage in order to harm a particular group
- Crackdowns on (perceived) dissent relating to the cause and/or management of environmental degradation or disasters
- Discriminatory denial of disaster relief

The legal doctrinal basis for recognition of refugee status in these kinds of cases is firmly established, reflecting common ground between narrower and broader interpretations of the refugee definition presented in Section 3.2. Cases representing each subcategory are briefly described in turn below.

3.6.1 The State Intentionally Causing Environmental Damage in Order to Harm a Particular Group

In fact no case was identified in which an individual actually sought recognition of refugee status in a situation where the state had intentionally caused environmental damage in order to harm a particular group. However, this kind of situation is identified a number of times in legal doctrine,[120] and is also recognised as potentially giving rise to refugee status by the NZIPT in the case of *AF (Kiribati)*,[121] which is considered in more detail in Section 3.7.

---

[120] See for example Walter Kälin and Nina Schrepfer, 'Protecting People Crossing Borders in the Context of Climate Change: Normative Gaps and Possible Approaches' 2012 UNHCR Legal and Protection Policy Research Series PPLA/2012/01, 31–32 <http://www.unhcr.org/4f33f1729.pdf> accessed 4 April 2019; McAdam, *Climate Change* (n 99); Bruce Burson, 'Environmentally Induced Displacement and the 1951 Refugee Convention: Pathways to Recognition' in T Afifi and J Jäger (eds), *Environment, Forced Migration and Social Vulnerability* (Springer 2010); Tracey King, 'Environmental Displacement: Coordinating Efforts to Find Solutions' (2005) 18 Geo Int'l Envt'l L Rev 543.
[121] *AF (Kiribati)* [2013] NZIPT 800413.

3.6.2 Crackdowns on (Perceived) Dissent Relating to the Causes and/or Management of Environmental Degradation or Disasters

Refugee Appeal No. 76374,[122] which concerned a claim for refugee status in New Zealand by a citizen of Myanmar (referred to in the determination by its other name Burma) who feared being persecuted as a consequence of her participation in humanitarian relief efforts in the aftermath of Cyclone Nargis in 2008, is commonly identified as an example of the kind of scenario where a person may establish a well-founded fear of being persecuted for a Convention reason in the context of 'natural' disasters and climate change.[123]

Refugee Appeal No. 76374 is a clear example of a case in which the disaster forms but the backdrop against which a classic pattern of state-led persecution for reasons of political opinion plays out. In her evidence, the claimant recounts a decades-long engagement in pro-democracy activism in Myanmar, including participation in demonstrations and assisting in the movement of dissidents within the country and transporting money on a few occasions. She was also involved in monitoring environmental degradation caused by mining activity.[124] It is within this context of political activism, and the claimant's affiliation with a particular political party, that her exposure to violence at the hands of the authorities arose in the aftermath of Cyclone Nargis.

The impact of Cyclone Nargis on Myanmar 'resulted in substantial injury and loss of life, destruction to property and infrastructure and causing internal population displacement'.[125] In the immediate aftermath of the cyclone, the claimant was contacted by her long-standing acquaintance (identified as BB) within the political party (identified as ABC) and asked to obtain funds that the party had sent to Myanmar. She used the money to buy food and materials for disaster relief, and arranged for their transport to the region most affected by the cyclone. She distributed food aid and provided funding to persons who had been orphaned as a result of the cyclone.[126] Before knowing that her conduct had aroused the interest of the authorities, she travelled to New Zealand, where BB was living. Whilst there she learned that the authorities had visited her business and told her assistant that they wanted to see the claimant. The claimant learned from her sister that the authorities had begun to arrest people who had assisted with the disaster relief effort in the wake of Cyclone Nargis.[127]

After accepting the claimant's evidence in its entirety, the Authority considers whether, on the facts as found, the appellant has a well-founded fear of being persecuted. Considering country of origin information relating to the imprisonment of people who had assisted in humanitarian relief in the aftermath of Cyclone Nargis, as well as general information about Myanmar's human rights record, the Authority concludes that, in light of her open

---

[122] *Refugee Appeal No. 76374* RSAA (28 October 2009).
[123] See for example Nansen Initiative, 'Agenda for the Protection of Cross-Border Displaced Persons in the Context of Disasters and Climate Change' (2015) <https://disasterdisplacement.org/the-platform/our-response> accessed 4 April 2019; Vikram Kolmannskog, 'Climate Change, Environmental Displacement and International Law' (2012) 24 J Int Dev 1071, 1076; McAdam, *Climate Change* (n 99) 50 [used here as an example of a claim arising in the context of a disaster, but where the claim itself is not really *about* the disaster]; Kälin and Schrepfer, 'Protecting People Crossing Borders in the Context of Climate Change' (n 120) 33; Bruce Burson, 'Protecting the Rights of People Displaced by Climate Change: Global Issues and Regional Perspectives' in Bruce Burson (ed), *Climate Change and Migration: South Pacific Perspectives* (Institute of Policy Studies 2010) 160.
[124] *Refugee Appeal No. 76374* (n 122) [14]. [125] Ibid [20]. [126] Ibid [127] Ibid [26].

involvement as the co-ordinator of the ABC Party's disaster relief efforts in the wake of Cyclone Nargis, the claimant 'would very likely [be] known to the regime' and that she 'faces a real chance of being sentenced to a substantial term of imprisonment as a result of an unfair trial process'.[128] In determining the claim, the Authority concludes that it has 'no doubt whatsoever that this appellant does have a well-founded fear of being persecuted if returned to Burma' and that 'the regime will in all probability impute a negative political opinion to the appellant for her independent facilitation of disaster-relief activity as it has done with others'.[129]

In light of the facts of the case, it is not surprising that *Refugee Appeal No. 76374* receives recognition as an example of the kinds of circumstances in which an individual may establish a well-founded fear of being persecuted for a Convention reason in the context of 'natural' disasters and climate change. Indeed, the case reflects a classic political refugee scenario that happens to play out against the backdrop of a disaster. There is no question about who the actor of persecution is, nor is there even a question about whether the actor of persecution is motivated to cause serious harm for a Convention reason.

Perhaps for this reason, other claims for recognition of refugee status in the context of Cyclone Nargis have also been successful. In *RRT Case No. 0903555*, the Refugee Review Tribunal of Australia (RRTA) found the claimant to be a refugee who had a well-founded fear of being persecuted for reasons of her membership of a particular social group of 'business people'.[130] The claimant had adduced credible evidence that she and her husband had supported political protests in 2007, were supporters of Aung San Suu Kyi, and had participated in the distribution of disaster relief in the aftermath of Cyclone Nargis in 2008. However, although the appeal was allowed with specific reference to the increased exposure of the claimant as a consequence of her involvement in the Cyclone Nargis relief effort,[131] the case ultimately turned on the likelihood that the claimant, as a businesswoman, would face extortion at the hands of the authorities.

A number of other cases concerned crackdowns by authorities on protests arising in the context of disasters. In *AL v Austria*, for example, a politically active man from Togo had sought asylum in Austria, expressing a fear of being killed by the authorities in his home country on account of his involvement in protests about the management of flood relief efforts in the country.[132] The Court summarises the basis of claim at paragraph 9 of its judgment:

> In summer 2008, areas of Togo were flooded and the applicant and his family lost their house and their possessions. They thus moved to a camp for flood victims. The applicant claimed that relief items sent by international organisations were unequally distributed in the camp, which was why he and a group of other young members of the UFC criticised the distribution methods and attempted to distribute the goods themselves. However, soldiers threatened the applicant by saying, inter alia, that after the 2010 elections, the applicant and other UFC party members 'would see what would happen'. This was the reason why the applicant decided to leave Togo and fled to Europe.[133]

---

[128] Ibid [40–41].
[129] Ibid [42] on well-founded fear of being persecuted and [45] on nexus to a Convention ground.
[130] *RRT Case No. 0903555* [2010] RRTA (15 January 2010).   [131] Ibid [61].
[132] *AL v Austria* (2012) ECHR 828.   [133] Ibid [9].

The applicant also recounted having been 'beaten up together with other protesters when they had organised a demonstration in the camp for flood victims'.[134] That case was dismissed by the Court because of changed political conditions in Togo.

In *YC v Holder*, a similar example of state violence in the context of disaster relief is recounted. In this case, the claimant 'asserted that he had been arrested, detained for 15 days, beaten and kicked in the stomach, and fined in June 2001 because he protested the local government's denial of disaster assistance after a typhoon destroyed his family's home and crops'.[135] However, the claim turned more expressly on the claimant's *sur place* pro-democracy political activity in the USA, and the claim was dismissed as not disclosing a risk on return.[136]

Finally, *RRT Case No. 1001325* concerned the fear held by a Chinese citizen that she would be arrested upon return because her husband had protested against the authorities' failure to prevent a flood, which had caused damage in their village, including the destruction of the couple's eel farm. The claim was dismissed on credibility grounds.[137]

In addition, a number of cases peripherally mentioned violence from non-state actors arising as a consequence of participation in disaster relief activities. In *RRT Case No. 060926579*, for example, members of Hindu organisations were angered by the fact that the claimant had been delivering disaster relief to Muslims.[138] The claim itself turned, however, on the claimant's general activities with the organisation and was ultimately dismissed on credibility grounds. Similarly, in *RRT Case No. 1104064*, an individual who claimed to have coordinated food relief for victims of the 2010 floods in Pakistan described how tensions had arisen between his organisation and some religious organisations that did not want his organisation to be collecting funds for the relief.[139] The central issue of this case, however, was the threat presented by the Taliban who, the claimant asserted, had targeted him for his involvement in organising entertainment activities.[140] This claim was also dismissed on credibility grounds.

From this survey it becomes clear that scenarios entailing the direct and intentional infliction of harm for a Convention reason will support a claim for recognition of refugee status, provided the account is found to be credible. In these cases, the unfolding of a disaster provides but a backdrop for familiar patterns of violence that present no difficulty for decision-makers.

3.6.3 Discriminatory Denial of Disaster Relief

Hathaway's 2014 intervention identified a number of cases relating to the discriminatory deprivation of disaster relief, predominantly used as a means to a political end.[141] Although

---

[134] Ibid [15].
[135] *YC v Holder, Attorney General*, 11-2749-ag, 11-3217-ag, United States Court of Appeals for the Second Circuit (18 December 2013) 10.
[136] Ibid 30–31. Note that the Court uses the 'clear probability' or 'more likely than not' standard under 8 CFR §1208.16(b)(2) in this case because, for technical reasons, although the claim was an asylum claim, the appeal was determined solely on the basis of the USA's withholding of removal provision.
[137] *RRT Case No. 1001325* [2010] RRTA 373 (11 May 2010).
[138] *RRT Case No. 060926579* [2007] RRTA 36 (28 February 2007).
[139] *RRT Case No. 1104064* [2011] RRTA 659 (3 August 2011) [21].   [140] Ibid
[141] Hathaway, 'Food Deprivation' (n 59) 327. See discussion in Chapter 1.

his focus is on how deprivation can support a claim for refugee status even in cases where the state 'simply could not be bothered'[142] to provide protection, the cases of food deprivation he cites relate entirely to scenarios in which the state intentionally deprives certain groups of disaster assistance for political reasons.[143] These cases are considered below.

Chan v Canada is of very limited relevance given that the case itself does not concern the denial of famine relief in anti-government areas. Rather, the case concerns a claim arising out of the enforcement of China's one child policy. Reference to denial of famine relief comes in a general discussion of the scope of 'the notion of "persecution"', where Desjardins J, citing an academic journal article, notes:

> Torture, beating, rape, are prime examples of 'persecution' but there are, presumably, a great many others. It has been suggested, for instance, that denial of famine relief in anti-government areas may come within that definition. The use of chemical warfare is another.[144]

Hagi-Mohammed v Minister for Immigration and Multicultural Affairs is also of limited assistance. Although the Federal Court acknowledges that '[o]ne can envisage circumstances in which the taking of harvests of those perceived as "enemies", rather than those perceived as "allies", might found a conclusion of persecution for a Convention reason',[145] the claimant's circumstances had nothing to do with such conduct.

RN (Returnees) Zimbabwe CG[146] is more directly concerned with the intentional deprivation of food than the preceding two cases, and the determination arguably makes a finding that the appellant has a well-founded fear of being persecuted for a Convention reason in the context, amongst other threats, of the discriminatory denial of food aid.

The appellant was a thirty-nine-year-old former teacher[147] from Matabeland South, close to the border with Botswana.[148] Focusing on changes in the situation in the country since a previous country guidance case,[149] including in relation to living conditions, the Tribunal expressly accepts 'that discriminatory exclusion from access to food aid is capable itself of constituting persecution for a reason recognised by the Convention'.[150]

The determination does not expressly find that the appellant, who the Tribunal recognises as a refugee,[151] has a well-founded fear of being persecuted for a Convention reason in the context of discriminatory deprivation of food aid, although the finding can be inferred from paragraph 269 when read with paragraphs 267 and 258.

---

[142] Ibid 336.
[143] *Chan v Canada* [1993] 42 ACWS 3d 259 ['denial of famine relief in anti-government areas']; *Hagi-Mohammed v Minister for Immigration and Multicultural Affairs* [2001] FCA 1156, [7] ['taking of harvests of those perceived as "enemies", rather than those perceived as allies']; *RN (Returnees) Zimbabwe CG* [2008] UKAIT 00083 [249] ['discriminatory exclusion from access to food is capable itself of constituting persecution'], cited in Hathaway, 'Food Deprivation' (n 59) 333.
[144] *Chan v Canada* (n 143) [69], citing Maureen Graves, 'From Definition to Exploration: Social Groups and Political Asylum Eligibility' (1989) 26 San Diego L Rev 740, 814.
[145] *Hagi-Mohammed* (n 143) [7].   [146] *RN (Returnees) Zimbabwe CG* (n 143).   [147] Ibid [7].
[148] Ibid [35].
[149] For a description of the country guidance system in the United Kingdom, see Sir Nicholas Blake, 'Luxembourg, Strasbourg and the National Court: The Emergence of a Country Guidance System for Refugee and Human Rights Protection' (2013) 25 IJRL 349.
[150] *RN (Returnees) Zimbabwe CG* (n 143) [249].   [151] Ibid [274].

At paragraph 269, the Tribunal finds that:

> This means, for the reasons set out above, that she has established a well-founded fear that she would be persecuted for a reason that is recognized by the Refugee Convention and that there is a real risk that she would be subjected to various forms of ill-treatment such as to infringe article 3 of the ECHR.[152]

One of the reasons supporting the finding is provided at paragraph 267 and concerns the appellant's ability to support herself by selling fruit and vegetables:

> Nor could the appellant seek to support herself in the only other way she had in the past, by selling fruit and vegetables with her mother. For someone in her position there are no such products left to sell and, in any event, her mother has fled the area on account of the threat from the militias arising because the area has been identified as one that supported the MDC in the elections.[153]

Earlier, at paragraph 258, the Tribunal finds:

> The fresh evidence now before the Tribunal demonstrates that the state is responsible for the displacement of large numbers of people so as to render them homeless and, unless the misgivings expressed in the evidence before us about the very recent lifting of the ban on the distribution of food aid prove to be unfounded, the evidence demonstrates also that there has been a discriminatory deprivation of access to food aid which, plainly, is a deliberate policy decision of the state acting through its chosen agents.[154]

As a former teacher from an area deemed loyal to the political opposition and with no family support, the appellant is seen by the Tribunal as a person who would face the discriminatory denial of food aid if returned to Zimbabwe. The Tribunal considers that such discriminatory denial amounts to persecution for a Convention reason. The determination is unequivocal in its recognition of the possibility that discriminatory denial of food aid can amount to persecution for a Convention reason and thus engender eligibility for refugee status.

Clearly significant in itself, the determination is also interesting for its presentation of the reasoning from the preceding country guidance case on Zimbabwe.[155] In that case, the Tribunal was also called upon to consider whether deprivation of food aid in the country engaged the United Kingdom's international protection obligations. However, the answer at the time *HS (returning asylum seekers) Zimbabwe* was determined, proved to be negative. The reasoning of the Tribunal in that case, reproduced in *RN*, reinforces the fundamental importance of identifying a culpable human agent in the context of a claim for recognition of refugee status, as reflected in the paradigm-level distinctions discussed in Chapter 1:

> Other than those who have been made homeless and displaced to areas where they have no support mechanisms to fall back on as a consequence of Operation Murambatsvina, citizens of Zimbabwe face such difficulties as they have to confront as a consequence of ill-judged political initiatives, economic mismanagement and unusually bad weather conditions affecting further the capacity of the country to produce crops.[156]

---

[152] Ibid [269]. [153] Ibid [267]. [154] Ibid [258].
[155] *HS (returning asylum seekers) Zimbabwe CG* [2007] UKAIT 00094.
[156] Ibid [60], cited in *RN (Returnees) Zimbabwe CG* (n 143) [254].

What made it possible for the Tribunal in *RN* to acknowledge the discriminatory deprivation of food aid contributed to the finding that the appellant had a well-founded fear of being persecuted for a Convention reason was the fact that the conduct was demonstrably direct and intentional, and thus falls within the readily recognisable scenarios reflected in Category 2.

Similar examples are found in the New Zealand jurisprudence. Although more focus is placed on the risk of exposure to physical violence, *Refugee Appeal No. 76237*[157] also recognises refugee status in a person in part because:

> Even if she is able to access money from her family abroad, food and other basic items are hard to come by. She will have no ready-made support networks to obtain items such as food which Zimbabweans are having to obtain from outside the country by a variety of informal and unreliable means. Much of the population is reliant now on food aid. This, however, is distributed discriminately in favour of ZANU-PF supporters. She would be unlikely to benefit from a distribution of food aid in these circumstances.[158]

The Tribunal expressly adopts a predicament approach to the nexus clause, considering that the claimant's gender (female), ethnicity (Indian), and political opinion are a *contributing cause* of her exposure to serious harm. Given the clear evidence that discriminatory denial of food aid was politically motivated in Zimbabwe, the RRTA had no difficulty recognising refugee status in a person who recounted personal experiences of being denied food aid because of her membership of the opposition MDC political party:

> On the totality of the evidence before it, the Tribunal is satisfied that in light of [Applicant 1]'s prior activism in the MDC, her continued membership in the MDC and her now lengthy absence from Zimbabwe, she would be at risk on her return of harassment, discrimination and the denial of basic services, including medical treatment, such as may threaten her capacity to subsist. The Tribunal is satisfied that such treatment would constitute persecution involving serious harm to [Applicant 1].[159]

Conditions in Zimbabwe remain in focus in *RS and Others (Zimbabwe – AIDS) Zimbabwe v Secretary of State for the Home Department*[160] with substantial attention being given to the discriminatory denial of food aid, including in the specific circumstances of the individual case. Unlike the cases considered so far within this subcategory, the determination in *RS and Others* reflects legal argument concerning the kinds of circumstances in which discriminatory denial of food aid can amount to persecution for a Convention reason.

The case concerned three claimants, all of whom had been diagnosed with HIV, and who had based their claims for international protection at least in part on difficulties accessing medical treatment for the condition. The determination dismisses each of the claims. The details of the individual claims, and the legal argument concerning eligibility for complementary protection under Article 3 ECHR, are not relevant for the purposes of this book, but the legal argument concerning discriminatory denial of food aid is.

The appellants relied on the recognition in *RN (Returnees) Zimbabwe* that discriminatory denial of food aid could amount to persecution for a Convention reason, giving rise to legal argument that is addressed from paragraph 91 of the determination:

---

[157] *Refugee Appeal No. 76237* RSAA (15 December 2008).   [158] Ibid [43].
[159] RRT Case No. 1002650 [2010] RRTA 595 (14 July 2010) [108].
[160] *RS and Others (Zimbabwe – AIDS) Zimbabwe v Secretary of State for the Home Department* CG [2010] UKUT 363.

91. Ms Grey referred to the evidence considered in RN of discriminatory exclusion and access to food aid and whether it was capable of amounting to persecution. There would need to be a careful examination of whether such deliberately discriminatory actions had had a real effect on the situation of any particular potential beneficiary and had led to a situation where a benefit was not received which otherwise would have been received. It was not enough simply to show some discrimination or skewing of what in any event were inadequate resources without also showing that the particular appellant would be adversely affected as a result.

92. In so far as it was argued that discriminatory access to food supplies or medical treatment amounted to persecution and was said to be as a result of the appellants' status as MDC supporters or their inability to demonstrate active support for ZANU-PF, lack of medical treatment or food due to insufficiency of resources would not give rise to a Convention breach. It would need to be shown that there was a deliberate withholding of food or medical treatment.[161]

As an argument advanced by the representative for the Secretary of State, it is unsurprising to see emphasis placed on the need to establish 'deliberate withholding of food or medical treatment', as distinct from the broader contributory cause approach adopted by the UNHCR, legal doctrinal scholars, and some courts.

Counsel for the appellants argued that:

> an arbitrary or deliberate denial of access to food or medical treatment on political grounds could amount to persecution and that discriminatory measures could constitute persecution if the consequences were sufficiently severe. The position was now confirmed by the provisions of the Qualification Directive and acts of persecution could take the form of administrative measures which were discriminatory or implemented in a discriminatory manner. If the appellants were able to show that there was a real risk that they would be denied medical treatment or food aid because they were not government supporters and that their suffering would be exacerbated for this reason then they would be entitled to refugee status.[162]

Arguing that arbitrary denial of access to food or medical treatment could also support a finding of 'persecution', Counsel for the appellants adopts a wider notion of 'persecution' than the representative for the Secretary of State.

The Tribunal clarifies the position it adopted in RN by articulating an expressly human-rights-based approach, tracing that approach in UK jurisprudence back to *Ullah v Special Adjudicator*[163] and Hathaway's definition of persecution as '[t]he sustained or systematic failure of state protection in relation to one of the core entitlements which has been recognised by the international community'[164] and accepts that 'the arbitrary or deliberate denial of access to food on political grounds may amount to persecution'.[165]

Noting on the basis of the substantial evidence it had considered, that the situation in Zimbabwe had not changed to such an extent as to warrant revisiting the country guidance in RN,[166] the Tribunal applies that guidance to the facts in the three cases before it. Having concluded that accounts of discrimination in access to medical treatment were 'anecdotal',[167] the Tribunal makes the same finding in relation to access to food,[168] concluding:

---

[161] Ibid [91–92].   [162] Ibid [105].   [163] *Ullah v Special Adjudicator* [2004] UKHL 26.
[164] *RS and Others (Zimbabwe – AIDS)* (n 160) [129].   [165] Ibid   [166] Ibid [199].   [167] Ibid [214].
[168] Ibid [215].

Bringing all this evidence together, we do not consider it has been shown that there is a real risk that any of the appellants would be denied food aid on grounds of political opinion. Certainly there is evidence of discriminatory denial of access to food, but we see that as being no more than sporadic and certainly not endemic. Nor do we consider that there is a real risk of harm to any of the appellants on the cumulative basis of access to medication and access to food.[169]

In relation to the individual applicants, each claim was dismissed on the basis that the claimant would be returning to opposition MDC party strongholds,[170] as well as on credibility grounds for some of the claimants.[171] The Tribunal also had regard to evidence that the ability of NGOs to distribute food aid had improved since *RN* was determined in 2008,[172] even though it did not consider the situation to be so different as to justify a departure from that guidance.

The determination is based on the specific circumstances of the individual applicants, but as a country guidance case concerning access for vulnerable individuals in situations of food insecurity to HIV treatment in the country, it also has more general application. It confirms that even in situations of food insecurity, the prospects of succeeding in a claim for refugee status on the basis of even the direct and intentional denial of food aid for political reasons can be limited. Although reinforcing the doctrinal position, expressly from a human-rights-based approach, that denial of food aid can amount to persecution, the Tribunal makes clear the importance of determining each claim in light of the personal circumstances of the individual and in light of the changing conditions in the country. The availability of relief from NGOs (which had been restricted when *RN* was determined),[173] for example, was a factor that affected the Tribunal's assessment.

Having considered these cases relating to the deliberate or arbitrary denial of food aid in Zimbabwe between 2008 and 2010, it is important to note, if only in passing, how international humanitarian organisations understood conditions on the ground. As Chapter 7 explains in more detail, the Integrated Phase Classification (IPC)[174] provides a snapshot of food insecurity in a particular country at a particular time and enables evidence-based projections into the future. In October 2008, the situation in most of Zimbabwe was classified as 'moderately food insecure',[175] while by the time *RS* was considered in 2010 the situation was classified as 'generally food secure'.[176] Without at this point describing the IPC, it is instructive to note that moderate food insecurity reflects a relatively low degree of food insecurity, below 'highly food insecure', 'extremely food insecure', and 'famine'.

Hence, although, as the determinations reflect, individuals were clearly facing difficulties accessing adequate food, including potentially to the point of experiencing denials of the right to an adequate standard of living under Article 11 ICESCR (especially as a

---

[169] Ibid [220].   [170] Ibid [295], [298], [300].   [171] Ibid [293].   [172] Ibid [144].
[173] *RN (Returnees) Zimbabwe CG* (n 143) [258].
[174] IPC Global Partners, 'Integrated Food Security Phase Classification Technical Manual Version 2.0: Evidence and Standards for Better Food Security Decisions' (Rome, FAO 2012).
[175] FEWS, Zimbabwe Food Security Outlook October 2008 to March 2009 <http://www.fews.net/sites/default/files/documents/reports/zimbabwe_outlook_2008_10.pdf> accessed 4 April 2019.
[176] FEWS, Zimbabwe Food Security Outlook January–June 2010 <http://www.fews.net/sites/default/files/documents/reports/Zimbabwe_Outlook_January_2010_final.pdf> accessed 4 April 2019.

consequence of the discriminatory nature of the denial, which could amount to a breach of Article 2(2) ICESCR), the Tribunal is also wise not to overstate the gravity of the situation.

3.7 Category 3: Other Failures of State Protection

Failures of state protection may be *ex post* a particular disaster event, such as when the state, in Hathaway's terms, 'could not be bothered' to protect a certain part of the population because of who they are. But legal doctrinal scholars have also anticipated the possibility of relevant conduct arising *ex ante* any particular disaster event, for instance in failures to invest in early warning mechanisms, failures to enforce building codes, failures to maintain flood protection systems, and so forth.[177] As the jurisprudence reveals, it can sometimes be difficult to distinguish cases along this *ex ante–ex post facto* continuum, as both scenarios may be reflected in the facts. Regrettably, many cases in this category do not reflect the 'anxious' or 'searching' scrutiny that each claim for international protection warrants, with relevant lines of enquiry not taken, at times owing to clear assumptions about the nature of 'natural disasters'. In what follows, cases reflecting *ex post* failures of protection are considered first, followed by cases reflecting *ex ante* failures. The final subcategory concerns cases where both *ex post* and *ex ante* failures are evident.

3.7.1 Ex post *Failures of State Protection*

Disasters may engender or exacerbate situations where individuals are exposed to acts of physical violence by family members, which the state is unwilling or unable to provide effective protection against. For example, it is widely recorded that disasters engender increases in sexual and gender-based violence.[178] It is therefore not surprising that at least one individual has found her predicament, entailing general post-disaster adversity combined with specific post-disaster intimate partner violence, to require international protection. In *Ferguson v Canada (Minister of Immigration and Citizenship)*,[179] a Grenadian woman and her children applied for recognition of refugee status in the aftermath of Hurricane Ivan, which destroyed their home and the primary applicant's business in September 2004. The claimant had travelled to Canada within one month of the hurricane making landfall in Grenada, and the initial basis of claim was the adversity to which she and the children were exposed as a consequence. That claim was rejected in July 2005, although the judgment under consideration does not provide details of the reasons for refusal.

---

[177] See in particular McAdam, *Climate Change* (n 99) 47–48.
[178] See for example Catarina Kinnvall and Helle Rydström (eds), *Climate Hazards, Disasters, and Gender Ramifications* (Routledge 2019); Rosalind Houghton et al, '"If There Was a Dire Emergency, We Never Would Have Been Able to Get in There": Domestic Violence Reporting and Disasters' (2010) 28 IJMED 270, citing a substantial number of studies confirming this phenomenon. See also John Twigg, 'Disaster Risk Reduction: Mitigation and Preparedness in Development and Emergency Programming', ODI Humanitarian Practice Network Good Practice Review No. 9 (March 2004) 82 <https://www.preventionweb.net/educational/view/8450> accessed 4 April 2019.
[179] *Ferguson v Canada (Minister of Citizenship and Immigration)* 2008 FC 903, 169 ACWS (3d) 629.

The claimant subsequently submitted a claim for permanent residence on humanitarian and compassionate grounds in August 2006 expressing a fear of being exposed to violence at the hands of her common law spouse. This claim was rejected by the decision-maker in May 2007 on the basis that there was sufficient state protection from spousal violence. It was the rejection of this claim that was the subject of the judicial review claim under discussion.

The claim was dismissed by the Ontario Federal Court on the basis that the decision-maker had made no reviewable errors of law. As the procedure did not entail a *de novo* determination of the claim, there is little in the determination to shed light on how epistemology and doctrine affect the determination of claims for recognition of refugee status in the context of 'natural' disasters and climate change. What the judgment does provide, however, is evidence that individuals can feel compelled to undertake long distance international travel in close temporal proximity to a disaster, and that such an individual may find cause to seek recognition of refugee status in that connection. It further highlights a scenario that does not feature prominently in legal doctrine, which relates to increased exposure to gender-based violence in the context of 'natural' disasters and climate change. The question whether the state is willing and able to provide effective protection from the conduct of non-state actors (including family members) in the aftermath of a disaster, will fall to be considered on a case by case basis applying established principles of international refugee law. The relevance of the disaster in this connection will concern the capacity of the state to provide protection, as well as, potentially, the evidence that spousal violence tends to increase in the aftermath of disasters.

A final significant observation arising from this case concerns the timeline. The claimants arrived in Canada within a month of the disaster and applied for recognition of refugee status. The claim was determined nine months later, by which time the conditions on the ground in Grenada would have changed significantly. The fact that relief work commences and the immediate dangers engendered by the disaster will tend to abate suggests that most claims for recognition of refugee status that rely on risks associated with the aftermath of a sudden-onset disaster will not succeed. However, the fact that the claimant and her children were able to remain in Canada while her claim was under consideration reflects the temporary protection inherent in the existing RSD system, so long as claims are not subjected to expedited procedures. The potentially narrow temporal scope of *ex post* disaster risk is acknowledged in Chapter 7 as a significant factor precluding recognition of refugee status in many, particularly more sudden-onset, scenarios.

Other cases raise questions about failures of state protection where the state may arguably have been in a position to provide protection, but appears unmoved to act. These cases differ from those considered in Category 2 as here the inaction does not result from an intention to cause harm, but by decisions about how scarce resources are to be distributed within a society. Such decisions may not be malign, but may reflect an inability or unwillingness on the part of the state to address the particular situation of vulnerable groups.

Refugee Appeal No. 70965[180] concerned the refusal of an application for recognition of refugee status by an Indo-Fijian man on the basis that his claim relating to adversity in the context of drought, generalised crime, and economic difficulties did not establish a well-

---

[180] *Refugee Appeal No. 70965/98* RSAA 289 (27 August 1998).

founded fear of being persecuted for a Convention reason. Although the Refugee Status Appeals Authority (RSAA) addresses elements of the claim relating to discrimination in the context of the right to work and the right to education, as well as concerns about violence, it does not address concerns about drought in any detail. This omission is puzzling given the fact that the Authority records the claimant as having submitted several newspaper articles referring to droughts and disaster, and advancing a claim grounded expressly in the failure of the state to protect Indo-Fijians from the impacts of the drought:

> Finally, on the topic of state protection, the appellant's representative has submitted that the drought which occurred in Fiji this year has left approximately 20% of the population (predominantly Indo Fijians) at 'destitution levels'. The appellant's representative goes on to submit that protection from starvation is a fundamental element of state protection and:
>
> '... because of a combination of natural disaster and socio-political unrest the government of Fiji is currently unable to afford such protection to the most vulnerable segment of its population, the land-less Indo-Fijian working class to which the appellant belongs'.

The response of the Authority to this claim is to conclude that:

> The Authority is of the view that the Fijian government is in effective control of its territory and has a police force and civil authorities that make serious efforts to protect its citizens from harm. This is clear from the DOS report and the newspaper articles submitted by the appellant in no way refute this conclusion.

Notwithstanding the claim that the appellant was differentially exposed, for a Convention reason, to disaster-related harm to the extent that drought had brought Indo-Fijians to 'destitution levels', this part of the claim is entirely ignored in the determination.

The RSAA rejects a similarly timed claim from an Indo-Fijian in *Refugee Appeal No. 70959/98*,[181] drawing on the same limited country of origin information and again failing to anxiously scrutinise the claim relating to the differential impact of drought in Fiji.

A much closer engagement with the question how the differential impacts of 'natural' disasters may support a claim for recognition of refugee status is provided in *RRT Case No. 071295385* ('the Sri Lanka tsunami case').[182] In this case the Refugee Review Tribunal considered whether an elderly Tamil woman without family support had a well-founded fear of being persecuted if returned to Sri Lanka in the aftermath of the 2004 Indian Ocean tsunami. Among other concerns, the claimant expressed a fear that she would face economic and social hardship, particularly as conditions of existence had become more difficult following the tsunami. The Tribunal accepted that the claimant would be vulnerable if returned to Sri Lanka, but did not accept that such vulnerability was sufficient to establish a claim for refugee status:

> The Tribunal accepts the evidence that life is very difficult for the applicant in Sri Lanka. It accepts the claims that it is difficult and expensive to get medical, pharmaceutical and ambulance assistance. It acknowledges the issues of poverty, immobility, isolation, loneliness, dependency, ill health and lack of nutrition affect the elderly. It accepts the claim that the applicant finds it difficult to survive on her monthly pension amount,

---

[181] *Refugee Appeal No. 70959/98* RSAA (27 August 1998).
[182] *RRT Case No. 071295385* [2007] RRTA 109 (20 June 2007).

particularly as the cost of living has risen following the tsunami. It accepts that as an elderly person, or an elderly woman or an elderly widow, the applicant is vulnerable in Sri Lanka. The independent information referred to above points to the difficulties in providing social services in Sri Lanka, particularly because of the drain on resources because of the on-going civil war and the effects of the 2004 Tsunami. It also identifies the elderly and elderly women in particular, as a vulnerable group in Sri Lanka. However, there are other vulnerable groups affected by these conditions as well. For example, there are orphans, youth, disabled people, and victims of the civil war and widows. These groups too would find life extremely hard in Sri Lanka with resources and government assistance limited. Being in a position of vulnerability or defencelessness will not of itself bring the harm feared within the Convention.[183]

The Tribunal refers to higher authority regarding the relevance of vulnerability to determining nexus to a Convention ground, as well as its own jurisprudence in support of its position. In *Omar Mohamud Mohamad v MIMA* [2000] FCA 109, the Federal Court of Australia had approved of the reasoning of the Tribunal that 'persecution of weak people in order to obtain what they have, because it is easier to recover what they have from them than from a stronger group is [not] persecution for a Convention reason'.[184] The Federal Court had in the first instance formed the view in *Mohamad* that

> [t]he persecution of small or weak groups is not per se persecution for a Convention reason. For a successful claim, an Applicant would need to establish that the relevant attacks were because the persecutors were aiming at destroying or damaging the members of the persecuted by reason of their membership of the particular clan.[185]

There is something of a chicken and egg problem within this approach, and the Federal Court does not ask itself why that particular group was weak in the first place. Of course, the Federal Court of Australia would not have been drawn into such consideration because of the statutory construction of the refugee definition, which only recognises refugee status where a person's race, religion, nationality, membership of a particular social group, or political opinion is 'at least the essential and significant reason'[186] for the persecution.

In the circumstances of this case, in which the post-tsunami adversity resulted from a weakened capacity on the part of the state to address the needs of all people within its jurisdiction, the claimant would not be able to show that her civil or political status (arguably the particular social group of elderly women) was what *motivated* the state to allocate insufficient resources to people in her predicament. A predicament approach, as summarised in Section 3.2, on the other hand, *would* recognise factors contributing to individual vulnerability as being relevant to the determination of refugee status. It would support the contention that a person may be persecuted in circumstances where the state simply cannot be bothered to protect people like the claimant, or makes decisions regarding the allocation of resources that, to the disadvantage of the claimant, are not based on objective and reasonable criteria. The predicament approach would similarly recognise a connection to a Convention reason where a person is differentially adversely affected by general failures of state protection in circumstances where her race, religion, nationality,

---

[183] Ibid
[184] Ibid, quoting *Mohamad v Minister for Immigration and Multicultural Affairs* [1999] FCA 688, [31].
[185] *Mohamad v Minister for Immigration and Multicultural Affairs* [1999] FCA 688, [30].
[186] s 91R(1)(a) of the Migration Act 1958 as it was at 20 June 2007.

membership of a particular social group, or political opinion is a contributory cause of such a predicament.

The Tribunal in the Sri Lankan tsunami case also draws upon case *MIMIA v VFAY* [2003] FCAFC 191, which concerns a young boy facing expulsion to Afghanistan. The Tribunal had concluded that, even if children or unaccompanied young people could be considered to constitute a particular social group, the risk he ran of being exposed to harm at the hands of the Taliban was not for reasons of his membership of a particular social group, but rather 'because of his youth and inexperience and so limited capacity to manage in a difficult environment and the generalised insecurity and hardship which prevails in his country'.[187] The Federal Court approved of the Tribunal's reasoning, observing that, in the paraphrasing of the Tribunal:

> The fact that the general conditions in Afghanistan might have a differential impact on some groups does not show that the members of those groups will be subject to persecution because of their membership of a particular social group.[188]

As with the *Omar Mohamud* case, the reasoning in *MIMIA v VFAY* [2003] FCAFC 191 reveals a limited willingness to look behind the vulnerability to the causes of such vulnerability. Clearly children have a certain innate vulnerability because they are children, as expressly recognised by the Committee on the Rights of the Child.[189] Hence, the fact that children are more susceptible to infectious disease, malnutrition, or exploitation does not automatically establish that children who suffer these forms of harm are persecuted. There must be an element of discrimination that permeates the notion of being persecuted for a Convention reason. But as Foster has demonstrated, this element is detectable in the way the state responds to the predicament of vulnerable groups.[190] Thus, doctrine suggests that a claimant's predicament can reflect the fact that she is persecuted for a Convention reason even when the actors of persecution do not have as their motivation the infliction of such harm for a Convention reason.

Having identified the legal test the claimant must satisfy as requiring evidence that 'the applicant faces a real chance of serious harm for the essential and significant reason of belonging to any of the particular social groups which the Tribunal has identified',[191] and making note of some efforts on the part of the Sri Lankan authorities to assist elderly people, the Tribunal concludes:

> The Tribunal does not accept that the essential and significant reason for inaction against the applicant would be that she is a member of a particular social group of *elderly people in Sri Lanka* or *elderly women in Sri Lanka* or *elderly widows in Sri Lanka*. The Tribunal therefore does not accept that any harm which might result if she were to return to Sri Lanka would be for the essential and significant reason of the applicant's membership of a particular social group or for any other Convention reason.[192]

---

[187] Ibid, quoting from the Tribunal determination that was subject of appeal in *MIMIA v VFAY* [2003] FCAFC 191.
[188] Ibid, paraphrasing *MIMIA v VFAY* (n 187), [60].
[189] CRC, 'General comment No. 13 (2011): The Right of the Child to Freedom from All Forms of Violence' (18 April 2011) CRC/C/GC/13, [72 (f)].
[190] See Foster, *International Refugee Law and Socio-Economic Rights* (n 37) 202 discussing *JDJ (re) No. A95-00633* CRDD No. 12, 28 January 1998.
[191] RRT Case No. 071295385 (n 182).   [192] Ibid

Although not unique in reformulating the clear wording of the Convention definition, it is worth considering here how significantly different the test articulated by the Tribunal is from the refugee definition at Article 1A(2) of the Refugee Convention. The claimant here is not required to establish a well-founded fear of being persecuted, but rather to establish that she faces a 'real chance of serious harm'. The difference between the Convention language and the alternative formulations that have developed in jurisprudence is addressed later in this book, where it is contended that such alternatives misconstrue the temporal and personal scope of the refugee definition.

3.7.2 Ex ante *Failures of Protection*

Under international human rights law, the state has a duty to respect, protect, and fulfil the rights of persons within the jurisdiction. Failure to take steps to prevent harm in the context of a foreseeable disaster can amount to a violation of human rights.[193] Only one case was identified in the survey that highlighted the responsibility of authorities for the harm to which the claimant feared being exposed if returned home. Although motivation-based approaches to the refugee definition would not recognise such unintentional, indirect causation of harm as supporting a claim for recognition of refugee status, such a claim may succeed where more human-rights-based interpretations prevail.

The claimant in *RRT Case No. N93/00894*[194] (the 'Bangladesh cyclone case') sought to rely on the Refugee Convention in the context of harm engendered by 'the ever-worsening natural disasters which characterise Bangladesh'. In this case, the claimant had arrived in Australia in 1988 and applied for recognition of refugee status in December 1989. The Refugee Review Tribunal recounts the evidence provided by the claimant, which depicts 'a wretched existence bordering on extreme poverty'. Periodic flooding exacerbated this predicament.

But the claimant did not contend that the predicament faced (predominantly by his family) was solely a consequence of the indiscriminate forces of nature. Rather, he sought to highlight the culpability of the Bangladeshi authorities:

> The Applicant claims that his poor economic circumstances are a direct result of the war of liberation in 1971 and the ever-worsening natural disasters which characterise Bangladesh. He claims that the devastating effect of the natural disasters can be directly attributed to a series of political decisions characteristic of ineffective government. These include the decision to log certain areas and to institute hydro-electric schemes on the border. This has meant that in heavy rains the devastation has been more severe.

If the facts were made out, this would be a clear example of a failure of state protection, where environmental degradation engendered by commercial and 'development' activity increases exposure to natural hazards. The claimant's representative asked the Tribunal to obtain independent reports in order to verify the applicant's claims in this connection, and also that his situation be considered on cumulative grounds.

In its determination the Tribunal ignores the claimant's request to obtain independent reports about natural disasters in Bangladesh, and concludes that he did not have a well-

---

[193] *Budayeva and Others v Russian Federation* [2014] 59 EHRR 2, *Öneryildiz v Turkey* [2005] 41 EHRR 20.
[194] *RRT Case N93/00894* [1996] RRTA 3244 (14 November 1996).

founded fear of being persecuted for a Convention reason. After concluding that the claimant's predicament did not reflect any form of discrimination, and forming the view that any adversity experienced in relation to employment, education, or access to healthcare would not amount to a violation of the claimant's rights under the International Covenant on Economic, Social and Cultural Rights (ICESCR), the Tribunal turns its attention to the phenomenon of 'natural disasters'. On this score the Tribunal concludes:

> The Tribunal accepts that natural disasters occur regularly and have a devastating impact on the population of Bangladesh. The Tribunal has every sympathy for the Applicant and his family in this regard. However again *there is no discriminatory intent* involved in these natural disasters. Refugee law exists to protect people from serious harm arising from an Applicant's civil or political status, therefore *people who fear harm from non-selective phenomenon* [sic], *such as earthquakes and floods, are excluded*. (See Hathaway, J. supra, p.93.) The Tribunal notes the Applicant's representative's submissions that the impact of the natural disasters has been worsened by bad political decisions. However again there has been no claim advanced that these bad political decisions were *deliberately* made so as to adversely impact on the Applicant, either as an individual or a member of a group of individuals, for one of the reasons specified in the Refugees Convention. Therefore this claim also falls outside the scope of the Refugees Convention and Protocol. [emphasis added]

As the italicised elements of the above paragraph highlight, the Tribunal's view reflects a very narrow approach to RSD. First, the requirement that any harm that befalls a person must be the result of the intentional, discriminatory infliction of harm by an individual actor of persecution is at variance with the more persuasive interpretation of the nexus clause reflected in the predicament approach discussed in Section 3.2. Second, the Tribunal clearly articulates assumptions inherent in the hazard paradigm, namely that earthquakes and floods are 'non-selective phenomena'. No consideration is given to the reasons why some people may be exposed and vulnerable to the impacts of such phenomena. Third, reliance on Hathaway reflects an incomplete reading of his work, taking only the general point about there tending not to be evidence of discrimination in disaster-related harm without reflecting his acknowledgement of the kinds of circumstances in which such a discriminatory context may be identifiable.

Although the claimant's request to the Tribunal to undertake country of origin research *proprio motu* may more appropriately have been directed to his own legal representative, who may have taken more detailed instructions about how the claimant's predicament was connected to one or more of the five Convention reasons, no examination of the political economy of disasters in Bangladesh is reflected in the determination at all.

3.7.3 Combined Ex post *and* Ex ante *Claims*

More careful scrutiny of claims relating to failures by the state to protect from the combined risk of earthquakes, tsunamis, and nuclear disaster is reflected in *RRT Case No. 1200203* ('the Australian Fukushima case').[195] The case before the Refugee Review Tribunal of Australia asserted that the Japanese government was engaged in persecutory conduct in

---

[195] *RRT Case No. 1200203* [2012] RRTA 145 (6 March 2012).

relation to its nuclear power policies and regulation, dissemination of information, and crisis management.

The claimant, a Japanese citizen, had travelled to Australia on a working holiday-maker visa in March 2010, one year before the accident at the Fukushima Daiichi Nuclear Plant following the earthquake and tsunami on 11 March 2011.[196] The Tribunal records the core of the claimant's fear at paragraph 27 of the determination:

> When asked what will happen to him if he returns to Japan the applicant said he will be killed by the radiation and the government's actions. When asked what action will kill him he said there are 55 nuclear plants in Japan and the government has not decided to close them. When asked why this will kill him he said there is evidence that the food and fish in Japan are contaminated. Also the government does not know how to dispose of the contaminated soil. He said the government has not provided instructions for those living in the affected areas. It has not provided measures for those living in the local areas so why would they provide instructions for those living far away in Osaka where he lives.[197]

Thus, the claimant articulates a fear of being exposed to serious harm as a result of both acts and omissions by state actors. The harm he expects being exposed to would arise immediately upon his return as well as gradually, as a result of exposure to radiation. As the claim relates to the consequences of a failure of the state to protect from the nuclear accident, it has an *ex post* element. There is also an *ex ante* element to the claim, as the claimant fears what he deems a foreseeable nuclear accident in the future, with similar consequences.

The case is significant in the study – seeking to determine the kinds of circumstances in which a person may establish a well-founded fear of being persecuted for a Convention reason in the context of 'natural' disasters and climate change – for two reasons. First, it raises the issue of when failures of state protection may engage the refugee definition. Second, it raises a question about the temporal scope of the refugee definition, given the claimant's expressed fear of being exposed to serious harm in the context of a potential future disaster. Although these two issues are closely interrelated, they will be discussed in turn below.

The application was refused in the first instance because the claimant 'did not actually claim that the government is actively seeking to harm him by deliberately killing, detaining, maiming, injuring or otherwise harming him'.[198] Further, the delegate did not accept that the harm the claimant feared 'fell within any of the grounds in the Convention'.[199] Again, as is by now familiar from much of the Australian jurisprudence surveyed, the idea that persecution entails deliberate infliction of serious harm significantly circumscribes the scope of refugee protection.

At the hearing, the claimant asserted that the fact that the authorities continue to 'push for nuclear power plants to be built' amounted to persecution.[200]

Applying Australia's statutory definition of 'persecution', the Tribunal asked the claimant:

> why he fears harm as a consequence of the disasters and nuclear accidents indicating in order to meet the definition of a refugee, the persecution that is feared has to be

---

[196] BBC, 'Japan Earthquake: Explosion at Fukushima Nuclear Plant' *BBC* (12 March 2011) <http://www.bbc.com/news/world-asia-pacific-12720219> 4 April 2019.
[197] *RRT Case No. 1200203* (n 195) [27].   [198] Ibid [32].   [199] Ibid   [200] Ibid [52].

motivated by one or more of the five reasons set out in the definition. It asked the applicant why the persecution feared involves conduct that is systematic and discriminatory.[201]

The Tribunal further asked, 'even if it was the case that there were more disasters, why it would be that the applicant was being singled out or targeted by the authorities or politicians'.

The Tribunal had regard to country information suggesting that 'the situation in Japan, outside the precautionary zone (80 kilometres from the Fukushima nuclear plant), has returned to normal'.[202] It further noted that 'the Japanese government is making efforts to test food for contamination and restrict the distribution of contaminated food and to assist the Japanese people and evacuate those in danger'.[203]

The claimant, who was legally represented, asserted a nexus between the harm feared and all five of the Convention grounds. The evidence provided by the claimant to the first-instance decision-maker is recounted by the Tribunal at paragraph 23 of the determination:

> As he is Japanese, born in Japan, he feels a victim of natural disasters and he has suffered because he lived there and his health was affected. He is not a Buddhist or Christian but his religion is that he trusted nuclear energy when he lived in Japan, and it was like a religion. He is a member of the particular social group that objects to living in a society with radiation in the air, due to the nuclear plants, and fears potential cancers. He has a political opinion regarding his objection to there being more than 55 nuclear plants in Japan, lots of radiation and contamination.[204]

On consultation with his legal representative, the claimant asserted that 'he was applying for a protection visa because the government is using nuclear plants as a nuclear weapon to kill the people in Japan'.[205]

The Tribunal dismissed the appeal, explaining:

> The Tribunal accepts that it might be that the applicant might be seriously harmed by a natural disaster such as an earthquake in the future if he were to return to Japan. However the Tribunal is not satisfied the applicant will suffer serious harm, which is systematic and discriminatory, on the basis of his race or nationality. It is not satisfied on the basis of the applicant's evidence that any harm he might suffer as a consequence of an earthquake would have an official quality or be tolerated officially. The Tribunal is satisfied on the basis of the independent information that the Japanese authorities have taken significant steps to assist those affected by earthquakes. It has not seen evidence that the applicant would be deprived of such assistance in the future if he were to be affected by an earthquake in Japan. The Tribunal does not accept the applicant will be persecuted on the basis of his race or nationality if he were to return to Japan.[206]

The Tribunal further does not accept that the claimant faces persecution for any other of the Convention reasons.[207]

It is accepted that 'the conduct leading to the harm is the government's nuclear energy policy and this could be considered systemic'.[208] However, the Tribunal continues, 'while the applicant may suffer serious harm it will not be persecution as it is missing a

---

[201] *RRT Case No. 1200203* (n 195) [56]. [202] Ibid [61]. [203] Ibid [68]. [204] Ibid [23].
[205] Ibid [31]. [206] *RRT Case No. 1200203* (n 195) [83]. [207] Ibid [84–92]. [208] Ibid [89].

discriminatory motivation by the Japanese government'.[209] On the specific concern relating to the state's management of its nuclear infrastructure, the Tribunal notes:

> The applicant has submitted the Japanese government is not able to manage its nuclear waste, it does not manage its nuclear power plants effectively and its old power plants are at risk. He believes as a consequence he will suffer harm. The Tribunal has not seen independent information to support the applicant's assertions. However even if the applicant is correct, there is no evidence before the Tribunal that a catastrophic consequence of these circumstances is intended to target and harm the applicant for a Convention reason. There is no evidence the applicant might suffer harm as the result of discriminatory conduct in relation to this claim.[210]

In dismissing the claim, the Tribunal concludes:

> Japan is a functioning democracy and the conduct relates to the governing of Japan. The conduct which the applicant claims will result in harm is not aimed at the applicant, or anyone in particular, but at governing Japan. *The conduct is not discriminatory* and any harm suffered as a consequence will not be persecution. Regardless of the harm being considered, as a consequence of governance, it will not be for a Convention reason.[211]

The Tribunal was not vexed by the claimant's reference to a potential future disaster as a source of harm relevant to the determination of his claim. However, it did frame its determination within a relatively narrow temporal scope. Setting out the law in the opening section of the determination, the Tribunal notes:

> Whether an applicant is a person to whom Australia has protection obligations is to be assessed upon the facts as they exist when the decision is made and requires a consideration of the matter in relation to the reasonably foreseeable future.[212]

The Tribunal provided the claimant with the opportunity to address concerns that his fear of being exposed to serious harm in the context of a potential future disaster may appear 'speculative', to which the claimant was only able to assert that Japan would be affected by many disasters in the future.[213] He did not adduce evidence regarding historical patterns of earthquakes, tsunamis, or nuclear accidents, nor did he adduce evidence of the future risk of such events unfolding. The Tribunal was clearly not moved to consider the temporal scope of the refugee definition on the basis of such unsupported claims, and framed its determination with reference to the 'reasonably foreseeable future', effectively rejecting the element of the claim relating to potential future disasters.[214]

This determination reflects many of the points identified throughout this book regarding the reasons why people fearing exposure to serious harm in the context of 'natural' disasters and climate change are generally unable to satisfy the requirements of the refugee definition at Article 1A(2) of the Refugee Convention. Specifically, the reasons identified by the Tribunal in this case include:

- The lack of intention on the part of the state to cause harm
- The indiscriminate impact of the disaster and the lack of evidence of any discriminatory motivation on the part of the state that resulted in the claimant being more exposed to serious harm than other people in Japan

---

[209] Ibid. [210] Ibid [97]. [211] *RRT Case No. 1200203* (n 195) [99] (emphasis added). [212] Ibid [18]. [213] Ibid [71]. [214] Ibid [82].

- The fact that the state was doing its best to address the consequences of the disaster
- The fact that the situation had largely returned to normal
- The lack of evidence that would make the claimant's fear of being exposed to serious harm in the context of a potential future disaster more than merely speculative

It is important to recognise that the claimant himself was unable to point to any element of discrimination that could establish a connection between his predicament and one or more of the five Convention reasons. This inability would prove fatal to the claim irrespective of the epistemological or doctrinal underpinnings of the determination.

Whilst focusing at times on the harm resulting from the earthquake in a way that echoes some of the generalising comments from senior courts surveyed in Chapter 1, the determination demonstrates a clear awareness of the relevance of the conduct of state actors in creating the conditions that engender a real risk of exposure to a nuclear accident in the event of a future natural hazard unfolding in Japan.

The Tribunal's focus (required by the statutory definition of persecution at section 91R (1) of the Migration Act 1958)[215] on the intention of the authorities is nevertheless noteworthy. It invites a far more narrow inquiry into the conduct of the state than a predicament approach would require. So long as the state did not intend for the accident to happen, and indeed intend for the accident to cause serious harm to the claimant for a Convention reason, it will be impossible for a person in the claimant's predicament to establish eligibility for refugee status. A human-rights-based approach, in contrast, would require examination of whether the accident, as well as *ex ante* and *ex post facto* conduct, revealed evidence of a state's failure to respect, protect, and fulfil rights guaranteed under the international human rights law, as a consequence of either acts, omissions, or both, and whether the consequences of such conduct amounted to a serious denial of human rights.[216] A connection to one or more of the five Convention reasons would also be required, but civil or political status need not be the sole or even predominant cause, but may merely be a contributory cause. Although the outcome of the claim may well have been the same, it is important at this stage to highlight differences in the methodology applied by different tribunals in determining claims for refugee status in the context of 'natural' disasters and climate change.

The Australian Fukushima case related to the risks arising from both earthquakes and nuclear power plants. The connection to state conduct was therefore clear, given the substantial governance issues surrounding the construction and operation of nuclear power plants. The case is interesting as it reflects in part what McAdam appears to have had in mind when identifying failures of disaster risk reduction as potentially being relevant to

---

[215] s 91R (now replaced with a similarly worded s 5J(4)) reads:

> For the purposes of the application of this Act and the regulations to a particular person, Article 1A(2) of the Refugees Convention as amended by the Refugees Protocol does not apply in relation to persecution for one or more of the reasons mentioned in that Article unless:
>
> (a) that reason is the essential and significant reason, or those reasons are the essential and significant reasons, for the persecution; and
> (b) the persecution involves serious harm to the person; and
> (c) the persecution involves systematic and discriminatory conduct.

[216] See Section 3.2 for discussion of the different approaches to the refugee definition.

RSD in the context of 'natural' disasters and climate change.[217] In this case, failures to ensure that facilities do not fail in the context of disasters, coupled with concerns expressed by the claimant about failures to inform the population of risks, are clear examples of *ex ante* conduct, which, if connected to a Convention reason, may support a claim for refugee status.

Two cases considered by the NZIPT provide further examples of how claims for international protection in the context of 'natural' disasters and climate change may give rise to considerations about both *ex post* as well as *ex ante* failures of state protection, including by raising questions about the temporal scope of the refugee definition. Both cases relate to a combination of slower-onset processes including sea-level rise and associated flooding and salination of agricultural land, together with concerns about the impact of more sudden-onset hazard events such as cyclones. The cases are considered together in order to consolidate the insights.

In *AF (Kiribati)*,[218] the NZIPT engages in the most carefully considered determination of a climate-change-related application for recognition of refugee status to date. The claim was brought by Mr Ioane Teitiota, a citizen of Kiribati. Mr Teitiota lived with his wife and three New Zealand-born children. Although they also faced being removed to Kiribati, they were not parties to the appeal.

In their evidence, the appellant[219] and his wife[220] expressed fear of the following forms of adversity and harm, arising from or related to sea level rise:

- Regular flooding of the land surface during high tides, affecting transportation networks
- Salination of drinking water
- Salination of the soil, destroying crops and making it difficult to grow new crops
- Damage to the sea wall in front of the appellant's parents-in-law's house
- Overcrowding
- Disease incidence
- Drowning in a tidal event or storm surge

In their evidence, the appellant and his wife acknowledged that 'the Government was taking what steps it could', including buying land on Fiji to grow crops.[221] The couple also accepted that the adversity they would be exposed to on return was 'common to people throughout Kiribati'.[222]

A second claim relating expressly to fears of being exposed to climate-change-related harm was considered by the NZIPT one year later. Here, the focus of the Tribunal fell clearly on the relevance of instruments extending complementary protection, with the Refugee Convention mentioned only insofar as to acknowledge that, in light of *AF (Kiribati)*, the claimants were not relying on that instrument in support of their claims for international protection.[223] Nonetheless, the case provides important doctrinal insights that are also relevant for the determination of refugee status, and it is in this connection that the determination is considered here.

---

[217] McAdam, *Climate Change* (n 99) 47–48. McAdam does not elaborate on the kinds of circumstances that could support a claim for recognition of refugee status in the context of failures of disaster risk reduction.
[218] *AF (Kiribati)* (n 121).   [219] Ibid [23–31].   [220] Ibid [32–33].   [221] Ibid [30].   [222] Ibid [31].
[223] *AC (Tuvalu)* [2014] NZIPT 800517-520 [45].

The claim was brought by four citizens of Tuvalu, who together comprised a family unit consisting of a wife, a husband, and their two children. The husband, who was born in the late 1970s, had worked as a teacher in Tuvalu until his departure for New Zealand in 2007. He had lived with his family in the family home, which was located near the community hall, medical facility, primary school, government office, church, and general store.[224] The husband also worked on different islands, during which time he rented accommodation. As a result of the death of his father in 2005 and an ensuing property dispute, the family home fell into disrepair.[225]

The husband described some of the adverse impacts of climate change that he had witnessed on Tuvalu, including the island where the family home was situated. Such impacts included:

- Partial submergence of land during monthly king tides
- Death of trees close to the shoreline
- Difficulties growing food
- Coastal erosion
- Water stress

In addition to these issues, the husband was concerned about difficulties obtaining teaching jobs on Tuvalu. He was particularly anxious about the potential risks his children would face from drinking contaminated and substandard water, which he considered had caused him to suffer from sores and skin complaints as a child.[226]

The wife, who was born in the early 1980s, had been educated at high school level and later worked as a pre-school teacher's aide. She had married the husband after they met whilst working at the same school.[227] As a result of two late term miscarriages, the wife was anxious about her children's future in Tuvalu, having regard to the lower standard of medical care available on the island compared with New Zealand. She was also concerned about the family's livelihood opportunities, as well as the lack of available accommodation.

Frequent inundation of the land by sea water makes plants difficult to grow and the houses are often surrounded by sea water during king tides.[228]

The Tribunal identifies a central issue to determine in the case:

> The central issue to be determined by the Tribunal is whether the government of Tuvalu can be said to be failing to take steps within its power to protect the appellants' lives from the effects of climate change such that their lives can be said to be 'in danger' and whether or not the harm they fear amounts to cruel treatment as that term is defined under the Act.[229]

Substantial country of origin information was considered by the Tribunal in both cases. In *AF (Kiribati)*, the Tribunal considered evidence from a country expert,[230] who explained his view that Kiribati was 'a society "in crisis" as a result of population pressure and climate change'.[231] Population pressure was said to be increasing social tension and causing freshwater to be depleted faster than it could be replenished. The lack of mainline sewage meant that increasing waste from the burgeoning population was contributing to contamination of fresh water resources. People are forced to rely heavily on fresh water delivery from the state utilities board, which is rationed. Sea level rise was resulting in increased salination

---

[224] Ibid [22].   [225] Ibid [25–25].   [226] Ibid [26–28].   [227] Ibid [29–30].   [228] Ibid [33].
[229] Ibid [3].   [230] *AF (Kiribati)* (n 121) [12–21].   [231] Ibid [13].

of soil and freshwater supplies, including as a result of more intense storms and associated storm surges. Vegetation has died in many areas, and some land has become uninhabitable because it is inundated with salt water three or four times per month. Salt water intrusion was said to be associated with an increased incidence of diarrhoea in children, which had resulted in some fatalities. The expert provided photographic evidence showing the impact of sea water inundation, including people whose possessions had been washed away.

In *AC (Tuvalu)*,[232] in addition to considering information submitted by the appellants' legal representative, the Tribunal also gathered sources of its own initiative and invited submissions from the appellants' representatives on the information. The determination itself does not set out the contents of the country of origin information, but does provide a list of the documents.[233] The documents include reports by the Office of the UN High Commissioner for Human Rights, reports to and by human rights monitoring bodies, including the Human Rights Committee and the Committee on the Rights of the Child, a statement by the UN Special Rapporteur on the Human Rights of Safe Drinking Water and Sanitation, a number of country profiles, including relating to climate change and water security, academic publications, including several publications by Professor Jane McAdam and empirical studies of human mobility in the region, and finally publications by the Tuvalu government.

In *AF (Kiribati)*, the appellant had argued in submissions that the harm he feared being exposed to upon return did not need to be attributed to the conduct of an identifiable actor of persecution, owing to the possibility of understanding the term in the passive voice as 'fleeing from something'.[234] This argument was rejected as reflecting the 'sociological conception' of refugee-hood, as distinct from the legal meaning.[235] First, the legal meaning of 'being persecuted' in New Zealand entailed the 'sustained or systemic violation of core human rights, demonstrative of a failure of state protection', a notion which the RSAA first adopted in *Refugee Appeal No. 11/91*,[236] articulated in detail in *Refugee Appeal No. 74665/03*,[237] and developed more fully in relation to the economic and social rights at issue in the appellant's case in *BG Fiji*.[238] Such a conception was predicated upon human agency, either by state or non-state actors.[239]

The Tribunal's explanation of how a human-rights-based approach to being persecuted would be applied in the context of 'natural' disasters and climate change warrants quoting at length:

> In the socio-economic sphere, although there is no right to a healthy environment contained in the International Convention [sic] on Economic, Social and Cultural Rights (ICESCR), Article 11 does provide a right to an adequate standard of living. Plainly, where natural disasters and environmental degradation occur with frequency and intensity, this can have an adverse effect on the standard of living of persons living in affected areas. Also relevant in this context is Article 12(1) which provides for a right 'to the highest attainable standard of physical and mental health'. In General Comment No 14 The Right to the Highest Attainable Standard of Health E/C12/2000 (11 August 2000)

---

[232] *AC (Tuvalu)* (n 223). [233] Ibid [34–36]. [234] *AF (Kiribati)* (n 121) [51].
[235] Ibid [52], citing Astri Suhrke, 'Environmental Degradation and Population Flows' (1994) 47 J Int Aff 473, 482.
[236] *Refugee Appeal No. 11/91* RSAA (5 September 1991).
[237] *Refugee Appeal No. 74665/03* RSAA (7 July 2004). [238] *BG (Fiji)* [2012] NZIPT 800091.
[239] *AF (Kiribati)* (n 121) [54].

at [14], the Committee on Economic Social and Cultural Rights stated that Article 12(2)(b) obliges states to take steps aimed at the prevention and reduction of the population's exposure to 'detrimental environmental conditions that directly or indirectly impact on human health'. In the context of the right to adequate food, the Committee has stated that 'States have a core obligation to take the necessary action to mitigate and alleviate hunger ... even in times of natural or other disasters' General Comment No 12 The Right to Adequate Food (Art11) E/C12/1999/5 (12 May 1999) at [6].

What these observations illustrate is that generalised assumptions about environmental change and natural disasters and the applicability of the Refugee Convention can be overstated. While in many cases the effects of environmental change and natural disasters will not bring affected persons within the scope of the Refugee Convention, no hard and fast rules or presumptions of non-applicability exist. Care must be taken to examine the particular features of the case.

... Nevertheless, like any other case, in cases where such issues form the backdrop to the claim, the claimant must still establish that they meet the legal criteria set out in Article 1A(2) of the Refugee Convention (or, for that matter, the relevant legal standards in the protected person jurisdiction). This involves an assessment not simply of whether there has been [a] breach of a human right in the past, but the assessment of a future risk of being persecuted. In the New Zealand context, the claimant's predicament must establish a real chance of a sustained or systemic violation of a core human right demonstrative of a failure of state protection which has sufficient nexus to a Convention ground.[240]

Building on these important observations, the Tribunal in *AF (Kiribati)* confirms that claims for international protection in the context of 'natural' disasters and climate change should be considered in the same manner and by applying the same principles as other claims relating to feared violations of economic and social rights. Of course, claims advanced in the context of 'natural' disasters may also entail fears of violations of civil and political rights as well.[241] Examples of conduct that could amount to persecution in this context are identified as including situations where:

- the provision of post-disaster humanitarian relief may become politicised
- where environmental degradation is used as a direct weapon of oppression against an entire section of the population, such as occurred with the Iraqi Marsh Arabs following the first Gulf War.[242]

Reference is also made at paragraph 69 to *Refugee Appeal No 76374* (the Cyclone Nargis case discussed in Section 3.6.2). The examples fall essentially within Category 2 of the taxonomy. There is no express recognition of the *ex ante* failures of disaster risk reduction identified by McAdam[243] but paragraph 62 in the section 'Environmental degradation and international human rights law' recognises 'it is now generally accepted that states have a primary responsibility to protect persons and property on their territory from the impact of disasters and to undertake disaster risk reduction measures'.

---

[240] Ibid [63–65].
[241] See for example *Refugee Appeal No. 76374* (n 122) relating to persecution by the Myanmar authorities of a person who assisted in the humanitarian relief effort following Cyclone Nargis in 2008.
[242] *AF (Kiribati)* (n 121) [58].   [243] McAdam, *Climate Change* (n 99) 47–48.

Although the specific scenarios identified do not reflect this, the Tribunal is expressly concerned to emphasise the possibility of refugee status arising in situations reflecting a failure of state protection, noting at paragraphs 54–55:

> The legal concept of 'being persecuted' rests on human agency. Although, historically, there has been variation in state practice on the issue, international refugee law has, in recent years, coalesced around the notion that it can emanate from the conduct of either state or non-state actors. In the former case, the failure of state protection derives from the inability of the state or lack of will to control its own agents who commit human rights violations, or its failure to take steps it is obliged to take under international human rights law. In the latter case, it derives from the failure of the state, via its human agents tasked with the requisite legal or regulatory authority, to take steps within their power to reduce the risk of harm being perpetrated by non-state actors.
>
> But this requirement of some form of human agency does not mean that environmental degradation, whether associated with climate change or not, can never create pathways into the Refugee Convention or protected person jurisdiction.

Hence, even though the specific scenarios identified by the Tribunal do not reflect *ex ante* failures of state protection, these are incorporated in principle.

In *AC (Tuvalu)*, the Tribunal further develops principles relevant to the determination of when a person has a well-founded fear of being persecuted for a Convention reason in the context of 'natural' disasters.

Significantly, the Tribunal engages in a detailed consideration of the ways in which the responsibility of the state to protect populations from serious disaster-related harm is engaged. Specific elements of the requirements under Articles 6 (right not to be arbitrarily deprived of life) and 7 (prohibition on torture or inhuman or degrading treatment or punishment) of the International Covenant on Civil and Political Rights (ICCPR) are not discussed here, as this book focuses exclusively on the Refugee Convention. However, a number of important observations made in relation to Articles 6 and 7 are relevant to RSD as well.

Just as the notion of persecution rests on human agency, so the requirement to establish a complementary protection claim under the ICCPR requires evidence of 'arbitrary' deprivation of life or ill-treatment, both of which are inconceivable without the identification of responsible human agency. Identification of relevant conduct giving rise to the harm feared is necessary in order to engage the *non-refoulement* obligation. Specific forms of 'treatment' are identified, including 'the discriminatory denial of available humanitarian relief and the arbitrary withholding of consent for necessary foreign humanitarian assistance',[244] the latter being closely examined in relation to what amounts to 'arbitrary' withholding of humanitarian assistance, having regard to General Comments from the Committee on Economic, Social and Cultural Rights, as well as the International Law Commission's Draft Articles on the Protection of Persons in Situations of Disaster, amongst other sources.[245] These forms of conduct will clearly also be relevant to determining failures of state protection in the context of RSD.

However, in the context of 'natural' disasters and climate change, it is important to be realistic about the capacity of the state to protect populations from natural hazards. Quoting

---

[244] *AC (Tuvalu)* (n 223) [84].   [245] Ibid [87–96].

at length from a passage in *Budayeva v Russian Federation*,[246] in which the Court identified the existence of a margin of appreciation in relation to states' positive obligations in the context of disaster risk reduction, the Tribunal considers that:

> Translating these statements into the context of the present appeals, this is not a case of a dangerous activity amenable to domestic regulation causing an environmental hazard due to poor regulation. The disasters that occur in Tuvalu derive from vulnerability to natural hazards such as droughts and hurricanes, and inundation due to sea-level rise and storm surges. The content of Tuvalu's positive obligations to take steps to protect the life of persons within its jurisdiction from such hazards must necessarily be shaped by this reality. While the Government of Tuvalu certainly has both obligations and capacity to take steps to reduce the risks from known environmental hazards, for example by undertaking *ex-ante* disaster risk reduction measures or though *ex-post* operational responses, it is simply not within the power of the Government of Tuvalu to mitigate the underlying environmental drivers of these hazards. To equate such inability with a failure of state protection goes too far. It places an impossible burden on a state.[247]

This observation is as important for RSD as it is for the determination of entitlement to complementary protection. States' capacity to address hazards, which may be increasing in frequency and intensity in the context of climate change, is limited. However, regard must be had to all of the circumstances of the case, including the nature and foreseeability of the particular hazard and the steps that the state may be expected to take, relying on its own resources as well as having recourse to international assistance.

AC (Tuvalu), therefore, appears as the first determination to expressly consider the possibility of international protection obligations arising in the context of failures of disaster risk reduction. Of course, although it is correct to recognise the limits of inhuman or degrading 'treatment' where a state is simply unable to provide protection in the context of climate change, an inability on the part of the state to provide effective protection does not rule out the possibility of recognition of refugee status, provided the necessary link to a Convention reason can be found in the claimant's predicament. This is an important distinction between refugee status and complementary protection in the context of disasters and climate change.

In *AF (Kiribati)*, the Tribunal only refers in passing to the 'well-founded' fear criterion of the refugee definition by noting that it requires evidence of a 'real, as opposed to a remote or speculative, chance of it occurring'.[248] The risk assessment is, however, taken up in more detail in relation to the complementary protection claim. In that connection, the Tribunal draws on the view expressed by the Human Rights Committee (HRC) in *Aaldersberg and Others v Netherlands*[249] (concerning the risk of being exposed to nuclear weapons) that the First Optional Protocol regulating the individual complaints mechanism required evidence of 'imminence' before it could be engaged. Here, the Committee had stated:

---

[246] *Budayeva v Russian Federation* (n 193) concerned the failure by the Russian Federation to protect the population of Tyrnauz from a mudslide, including by failing to have adequate disaster risk reduction systems in place.
[247] *AC (Tuvalu)* (n 223) [75].   [248] *AF (Kiribati)* (n 121) [53], following *Chan* (n 11).
[249] *Aaldersberg and Others v Netherlands* CCPR/C /87/D/1440/2005 (14 August 2006).

> For a person to claim to be a victim of a violation of a right protected by the Covenant, he or she must show either that an act or an omission of a State party has already adversely affected his or her enjoyment of such right, or that such an effect is imminent, for example on the basis of existing law and/or judicial or administrative decision or practice.[250]

The Tribunal is careful, however, not to misconstrue a higher standard of proof than what is actually required in order to engage the *non-refoulement* obligation under Articles 6 and 7 ICCPR:

> This requires no more than sufficient evidence to establish substantial grounds for believing the appellant would be in danger. In other words, these standards should be seen as largely synonymous requiring something akin to the refugee 'real chance' standard.[251]

Bruce Burson, the NZIPT Member who drafted both *AF (Kiribati)* and *AC (Tuvalu)*, has since commented that he would not have formulated the relationship between the imminence standard and the well-founded fear standard in this way:

> In retrospect, I would phrase the decision slightly differently as I believe it is a mistake to import even the language of imminence into RSD. Whatever its merits for other branches of international law, it is fundamentally ill-suited to the task of RSD, particularly once the well-understood evidential handicaps typically faced by claimants is acknowledged. It is a far more onerous task for the claimant to discharge the evidential burden typically resting on him or her to establish the claim if imminent risk has to be established.[252]

How RSD should address temporal challenges arising in the context of 'natural' disasters and climate change is taken up in Chapter 5.

The Tribunal in *AC (Tuvalu)* articulated additional legal principles relevant to understanding the temporal scope of the refugee definition. In relation to the question of risk, the Tribunal confirmed the approach it had taken in *AF (Kiribati)* that 'the evidence must establish that a future risk amounting to more than mere conjecture or surmise, but need not be established as being "more probable than not" or "likely to occur"'.[253] The Tribunal then articulates a guiding principle:

> This forward looking assessment of risk means that the slow-onset nature of some of the impacts of climate change such as sea-level-rise will need to be factored into the inquiry as to whether such 'danger' exists at the time the determination has to be made. As to whether anticipated harm arising in the context of slow-onset processes may reach the threshold of the claimant being 'in danger', much will depend on the nature of the process in question, the extent to which the negative impacts of that process are already manifesting, and the anticipated consequences for the individual claimant. The assessment is necessarily context dependent.[254]

---

[250] Ibid [6.3], cited in *AF (Kiribati)* (n 121) [89].   [251] *AF (Kiribati)* (n 121) [90].

[252] Bruce Burson, 'The Concept of Time and the Assessment of Risk in Refugee Status Determination' (Presentation to Kaldor Centre Annual Conference, 18 November 2016) <http://www.kaldorcentre.unsw.edu.au/sites/default/files/B_Burson_2016_Kaldor_Centre_Annual_Conference.pdf> accessed 5 April 2019. An argument that there is no imminence standard in international refugee law is advanced in Matthew Scott, 'Finding Agency in Adversity: Applying the Refugee Convention in the Context of Disasters and Climate Change' (2016) 35 Refug Surv Q 26, 55.

[253] *AC (Tuvalu)* (n 223) [57].   [254] Ibid [58].

The Tribunal, building on its observations in *BG (Fiji)*,[255] takes the opportunity to expand upon the range of considerations that will be relevant to the conduct of the risk assessment:

> Just as in the refugee context past persecution can be a powerful indicator of the risk of future persecution, so too can the existence of a historical failure to discharge positive duties to protect against known environmental hazards be a similar indicator in the protected person jurisdiction. Nevertheless, given the forward looking nature of the inquiry, the nature of the hazard, including its intensity and frequency, as well as any positive changes in disaster risk reduction and operational responses in the country of origin, or improvements in its adaptive capacity, will need to be accounted for.[256]

These observations clearly reflect a serious attempt by the NZIPT to address fears expressed by individuals that they will be exposed to serious harm in the context of a foreseeable future disaster or as a consequence of other climate-related impacts, such as sea level rise. The Tribunal draws on the doctrine that past persecution is indicative of future risk, but also notes that multiple factors can intervene between a previous exposure to serious harm and a potential future exposure. There is significant uncertainty in respect of both the scale of future climate impacts, as well as the steps that states may take to protect populations within their jurisdiction. Noting that the assessment is 'necessarily context dependent' is appropriate. However, the Tribunal provides no further practical guidance on how the assessment of risk in the context of a foreseeable future disaster or other adverse impacts of climate change may be conducted. Chapter 7 examines some possible sources of evidence and considers them in a demonstration of how the recalibrated human-rights-based interpretation of the refugee definition developed in Chapters 5 and 6 would be applied to claims for recognition of refugee status in the context of disasters and climate change.

The Tribunal in *AF (Kiribati)* does not set out the law relating to the nexus clause, but rather notes at paragraph 65 that there must be 'a sufficient nexus to a Convention ground'. Importantly, referring to paradigm-level assumptions, the Tribunal does acknowledge at paragraphs 56–57 that whilst

> the effects of natural disasters are often felt indiscriminately ... broad generalisations about natural disasters and protection regimes mask a more complex reality. The relationship between natural disasters, environmental degradation, and human vulnerability to those disasters and degradation is complex. It is within this complexity that pathways can, in some circumstances, be created into international protection regimes, including Convention-based recognition.

As the claimant in *AC (Tuvalu)* did not advance a claim under the Refugee Convention, nexus was not considered in the determination.

In *AF (Kiribati)*, the Tribunal held that what the appellant feared he would be exposed to if returned to Kiribati would not amount to being persecuted within the meaning attached to the notion under New Zealand law. Additionally, the Tribunal held that the cumulative impact of the difficulties that the Appellant would face on return to Kiribati would not amount to serious harm for the purposes of the Refugee Convention. The nexus requirement would also not be satisfied, given the evidence that the harm was faced by the population on Kiribati generally, indicating a lack of connection between the harm feared and one or more of the Convention reasons. There was also no evidence that the

---

[255] *BG (Fiji)* (n 238) [98].   [256] *AC (Tuvalu)* (n 223) [69].

government of Kiribati had failed to take steps to protect the population for a Convention reason. Consequently, the Tribunal determined that the claimant was not a refugee. The disposal of the claim for refugee status does not address whether the appellant's fear of being exposed to serious harm on return to Kiribati was well-founded. In part, this is understandable because the Tribunal accepted that the conditions the appellant feared being exposed to upon return had been objectively established. They were simply not indicative of a well-founded fear of being persecuted for a Convention reason. The only fear that entailed an element of risk assessment was the concern expressed by the appellant's wife that her children may be swept away in a storm surge. This part of the claim was, however, addressed in relation to the claim for protected person status, relating to Articles 6 and 7 of the ICCPR.

In relation to the claim under Article 6 of the ICCPR, the Tribunal considers the evidence that the government of Kiribati was actively taking steps to protect its population from the adversity identified by the claimants and more generally in evidence. For this reason, the Tribunal finds no evidence of a risk of *arbitrary deprivation of life*. Additionally, the Tribunal notes the lack of evidence of any imminent risk of harm as underpinning the determination.

In forming the view that neither Article 6 nor Article 7 of the ICCPR are engaged, the Tribunal emphasises the lack of evidence of any imminent risk of harm awaiting the appellant and his family should they return to Kiribati:

> The Tribunal accepts that, given the greater predictability of the climate system, the risk to the appellant and his family from sea-level-rise and other natural disasters may, in a broad sense, be regarded as more 'imminent' than the risk posed to the life of complainants in *Aaldersberg*. Nevertheless, the risk to the appellant and his family still falls well short of the threshold required to establish substantial grounds for believing that they would be in danger of arbitrary deprivation of life within the scope of Article 6. It remains firmly in the realm of conjecture or surmise... The Tribunal notes the fear of the wife in particular that the young children could be drowned in a tidal event or storm surge. No evidence has been provided to establish that deaths from these events are occurring with such regularity as to raise the prospect of death occurring to the appellant or his family member to a level which rises beyond conjecture and surmise at all, let alone a risk which can be characterised as an arbitrary deprivation of life in the sense outlined above.[257]

Thus, Mr Ioane Teitiota had failed to establish a well-founded fear of being persecuted on the basis of current conditions prevailing in Kiribati, and there was no imminent risk of being exposed to serious harm upon return. Potential future risks relating to climate change did not go beyond mere conjecture or surmise.

Mr Teitiota appealed to the New Zealand High Court, which heard the case as *Teitiota v The Chief Executive of the Ministry of Business Innovation and Employment*.[258] On the subject of persecution, the claimant now argued that climate change was an indirect but worldwide human agency, resulting from two centuries of greenhouse gas emissions, and individuals facing exposure to climate-change-related harm were thus entitled to protection under the Refugee Convention. Priestly J concludes that Mr Teitiota, or others in similar circumstances, could not bring himself within the scope of the Refugee Convention. At paragraph 11, he notes:

---

[257] *AF (Kiribati)* (n 121) [91].
[258] *Teitiota v The Chief Executive of the Ministry of Business Innovation and Employment* [2013] NZHC 3125.

it is abundantly clear that the displacement of such refugees has not been caused by persecution. Nor, importantly, have they become refugees because of persecution on one of the five stipulated Refugee Convention grounds. A person who becomes a refugee because of an earthquake or growing aridity of agricultural land cannot possibly argue, for that reason alone, that he or she is being persecuted for reasons of religion, nationality, political opinion, or membership of a particular social group.

Although his use of the qualifier 'for that reason alone' can be taken to recognise the range of scenarios identified by the NZIPT as reflective of the kind of human agency necessary to support a claim for recognition of refugee status, Priestly J's judgment here is more reminiscent of the hazard paradigm approaches reflected in the opinions of other senior judges surveyed in Chapter 1. Without appreciating the nuance and detail of *AF (Kiribati)*, there is a risk that decision-makers reading Priestly J's judgment will find reinforcement of the views expressed inter alia in *Applicant A and Another*, *Horvath*, and *Ward*, with potentially significant implications for the determination of refugee status in the context of 'natural' disasters and climate change.

This concern is reinforced by the nearly unequivocal rejection of claims brought by 'economic and environmental migrants':

> Humanitarian concerns and the issues of economic and environmental migrants or refugees are topics which individual states in the international community generally have to consider. But the Refugee Convention is not an available avenue for such migrants and refugees. Certainly it is not available to this applicant and his family.[259]

The explicit public policy considerations are made clear at paragraph 51:

> On a broad level, were they [i.e. the counsel's submissions regarding climate change impacts amounting to persecution] to succeed and be adopted in other jurisdictions, at a stroke, millions of people who are facing medium-term economic deprivation, or the immediate consequences of natural disasters or warfare, or indeed presumptive hardships caused by climate change, would be entitled to protection under the Refugee Convention or under the ICCPR. It is not for the High Court of New Zealand to alter the scope of the Refugee Convention in that regard. Rather that is the task, if they so choose, of the legislatures of sovereign states.[260]

Leave to appeal against the determination of the Tribunal was refused, with the conclusion at paragraph 64 that '[t]he attempt to expand dramatically the scope of the Refugee Convention and particularly Article 1A(2) is impermissible'.

Further closing the door to people seeking recognition of refugee status in the context of 'natural' disasters and climate change, the New Zealand Court of Appeal dismissed Mr Teitiota's application for leave to appeal to the Court of Appeal, noting at paragraph 41:

> No-one should read this judgment as downplaying the importance of climate change. It is a major and growing concern for the international community. The point this judgment makes is that climate change and its effect on countries like Kiribati is not appropriately addressed under the Refugee Convention.[261]

It goes without saying that this unequivocal statement, looking like many others surveyed in Chapter 1 to the hazard itself and its effect on the majority, fails to appreciate the social

---

[259] Ibid [44].   [260] Ibid [51].
[261] *Teitiota v The Chief Executive of the Ministry of Business Innovation and Employment* [2014] NZCA 173, [41].

context in which a minority may establish a well-founded fear of being persecuted for a Convention reason. *AF (Kiribati)* was not at all as unequivocal.

A more balanced conclusion is reached by the Supreme Court, which concludes:

> [12] We agree with the Courts below that, in the particular factual context of this case (even with the addition of the new evidence), the questions identified raise no arguable question of law of general or public importance. In relation to the Refugee Convention, while Kiribati undoubtedly faces challenges, Mr Teitiota does not, if returned, face 'serious harm' and there is no evidence that the Government of Kiribati is failing to take steps to protect its citizens from the effects of environmental degradation to the extent that it can.
>
> [13] That said, we note that both the Tribunal and the High Court emphasised their decisions did not mean that environmental degradation resulting from climate change or other natural disasters could never create a pathway into the Refugee Convention or protected person jurisdiction. Our decision in this case should not be taken as ruling out that possibility in an appropriate case.[262]

In *AC (Tuvalu)* the Tribunal does not accept that reliance by the government of Tuvalu on international assistance, for example relating to the provision of fresh water, exposed the claimant to a risk of *arbitrary* deprivation of life or cruel, inhuman, or degrading *treatment*.[263]

In relation to the Article 6 ICCPR claim, the Tribunal concludes:

> it has not been established that Tuvalu, as a state, has failed or is failing to take steps to protect the lives of its citizens from known environmental hazards such that any of the appellants would be in danger of being arbitrarily deprived of their lives.[264]

In relation to the Article 7 ICCPR claim, the Tribunal concludes:

> On the evidence before the Tribunal, there is no basis for finding that there is a danger of the appellants being subjected to cruel treatment by the state failing to discharge its obligations to protect its population and territory from the adverse impacts of natural disasters and climate change.

Thus, a determination regarding whether the risk of death or serious harm would amount to *arbitrary deprivation of life* or cruel, inhuman, or degrading *treatment* must necessarily consider the manner in which the authorities in the receiving state have addressed disaster risks, but must also take into account the nature of the risk and the ability of the state to address them. As noted earlier, the same does not necessarily hold in the context of RSD, where a person may be persecuted in situations where the state, despite its best efforts, simply cannot provide effective protection. This may not amount to inhuman or degrading treatment under complementary protection principles, but it does not rule out a finding that a person is persecuted in this connection, provided the nexus to a Convention reason is established.

At the same time as the NZIPT dismissed the international protection claim in *AC (Tuvalu)*, it granted the humanitarian claim in a separate determination under the case

---

[262] *Teitiota v The Chief Executive of the Ministry of Business, Innovation and Employment*, [2015] NZSC 107.

[263] *AC (Tuvalu)* (n 223) [62].   [264] Ibid [108].

name *AD (Tuvalu)*.²⁶⁵ The determination applied section 207 of the Immigration Act (2009), which reads:

> (1) The Tribunal must allow an appeal against liability for deportation on humanitarian grounds only where it is satisfied that
>
> (a) There are exceptional circumstances of a humanitarian nature that would make it unjust or unduly harsh for the appellant to be deported from New Zealand; and
> (b) It would not in all the circumstances be contrary to the public interest to allow the appellant to remain in New Zealand.²⁶⁶

Applying this provision, the Tribunal considers the presence of numerous family members in New Zealand, the claimants' wider ties to the community, including through the children's school and their involvement in the church, and the best interests of the children. This latter consideration is expressly based on Article 3 of the Convention on the Rights of the Child 1989, and the assessment places emphasis on the fact that the children had never been to Tuvalu, the fact that the eldest child had started school and, of significance for future cases relating to international protection in the context of 'natural' disasters and climate change, the fact that 'their young age makes them inherently more vulnerable to natural disasters and the adverse impact of climate change' that had been considered in the case.²⁶⁷ By recognising that the determination of the children's status in New Zealand necessarily requires a best interests assessment under Article 3 CRC, the NZIPT reinforces doctrinal arguments concerning the significant power of the CRC in protecting children from removal to countries where they risk being exposed to serious harm.²⁶⁸

Claims based on a fear of being exposed to serious harm in the context of 'natural' disasters and climate change continue to be lodged in New Zealand, and the principles articulated by the NZIPT in *AF (Kiribati)* have been acknowledged.²⁶⁹ In *AV (Nepal)*,²⁷⁰ a married couple left Nepal and applied for international protection approximately one year after the April 2015 earthquake in that country. The determination does not reflect any enquiry into the role of the claimants' race, religion, nationality, membership of a particular social group, or political opinion in contributing to their exposure to disaster-related harm, not least because this element of the refugee definition is only considered *after* a determination has been reached concerning the question of whether the claimants have a well-founded fear of being persecuted. In relation to this element, the Tribunal concludes:

> The appellants have not presented any evidence to support the notion that, if they were returned to their home in Kathmandu, there is a real chance that they will suffer serious physical harm arising from the sustained or systemic violation of internationally recognised human rights, demonstrative of a failure of state protection (see [30] above). Whatever harm the appellants may face in Nepal does not arise from the legal concept

---

²⁶⁵ *AD (Tuvalu)* [2014] NZIPT 501370-371.   ²⁶⁶ Ibid [17].   ²⁶⁷ Ibid [25].
²⁶⁸ See Guy S. Goodwin-Gill, 'Unaccompanied Refugee Minors: The Role and Place of International Law in the Pursuit of Durable Solutions' (1995) 3 Int'l J Child Rts 405; Jane McAdam, 'Seeking Asylum under the Convention on the Rights of the Child: A Case for Complementary Protection' (2006) 14 Int'l J Child Rts 251; Pobjoy, *The Child in International Refugee Law* (n 63).
²⁶⁹ See for example *AF (Tuvalu)* [2015] NZIPT 800859.   ²⁷⁰ *AV (Nepal)* [2017] NZIPT 801125-126.

of 'being persecuted' which requires human agency or omission (subject to the points made in the careful discussion in AF (Kiribati) [2013] NZIPT 800413 at [55] et seq).[271]

The determination provides further evidence that persons displaced in the context of 'natural' disasters and climate change continue to seek recognition of refugee status owing to a fear of being exposed to serious harm on return. It further confirms that those people who seek recognition of refugee status tend not to be among those who may actually establish a well-founded fear of being persecuted for a Convention reason in this connection. The determination is also interesting because, in contrast to the very careful consideration of country of origin information reflected in *AF (Kirbiati)* and *AC (Tuvalu)*, the conclusion that there is no evidence that the claimants' predicament on return would entail exposure to 'serious violations of internationally recognised human rights demonstrative of a failure of state protection' is based on three reports, summarised at paragraph 33:

> The April 2015 Nepal earthquake killed nearly 9,000 people and injured nearly 22,000. It had a magnitude of 7.8 Mw (United States Geological Survey, 25 November 2015) and it was the worst natural disaster to strike Nepal since the 1934 Nepal-Bihar earthquake ('Nepal earthquake: Eerie reminder of 1934 tragedy' The Economic Times (25 April 2015)). In 2015, hundreds of thousands of people were made homeless across many districts of the country and centuries old UNESCO world heritage sites in the Kathmandu valley were destroyed. There were continued aftershocks, including one on 12 May 2015 with a magnitude of 7.3 Mw, which killed more than 200 people and injured more than 2,500 ('Nepal earthquake: Dozens die in new tremor near Everest' BBC (12 May 2015)).[272]

There is no consideration of the steps that the state took to reduce the risk of exposure to harm in the context of the earthquake, nor examination of disaster response beyond noting that the claimants were 'provided with basic shelter and food immediately after the earthquake'.[273] The risk of exposure to serious harm in the context of a foreseeable future earthquake is not addressed. How different groups of people within Nepal were and continue to be affected by the earthquake remains unexplored. The limited information about the predicament of the claimants does not strongly suggest that the claimants could have established a well-founded fear of being persecuted for a Convention reason in the context of the 2015 earthquake or a foreseeable future disaster, but as one of the first cases to follow on from the path-breaking *AF (Kiribati)* and *AC (Tuvalu)*, the determination is disappointing in its failure to actually apply the careful methodology articulated by the Tribunal in those cases. Moreover, reference to the hundreds of thousands made homeless invokes the narrative of indiscriminate harm due to the forces of nature, which the Tribunal in *AF (Kiribati)* was at pains to transcend. The language of earlier determinations is revived in the concluding paragraph dismissing the claim for recognition of refugee status:

> Finally, the harm faced as the result of the 2015 earthquake, its shocks and any future seismic activity is faced by the population of Kathmandu and/or Nepal generally. Any potential danger does not arise by reason of the appellants' race, religion, nationality, membership of any particular social group or political opinion. Therefore, not only is there an absence of persecution, but also an absence of a Refugee Convention reason.[274]

---

[271] Ibid [36].   [272] *AV (Nepal)* (n 270) [33].   [273] Ibid [37].   [274] Ibid [38].

3.8 Unresolved Doctrinal Issues Arising from the Review

The hazard paradigm continues to exercise a strong grip on RSD, even in the jurisdiction that has produced the most clear and principled approach to date.

The jurisprudence reviewed in this chapter reflects the continued dominance of the hazard paradigm, even though more recent determinations by the NZIPT takes decisive steps towards a more social paradigm. The review confirms legal doctrine pointing to the availability of refugee status for persons who face direct and intentional infliction of serious harm for a Convention reason in the context of disasters and climate change, as such cases can be accommodated within interpretations of the refugee definition that make motive critical, as well as those adopting a more predicament-based approach. The prospects appear far more narrow for individuals pointing to other failures of state protection, whether *ex post* or *ex ante* any particular disaster. In jurisdictions where motive is critical, *ex ante* or *ex post* failures of protection that are not intentional have no prospect of success. Where a human-rights-based predicament approach is adopted, such as in the NZIPT jurisprudence, a narrow window opens for claims where the state fails to fulfil disaster risk reduction obligations. When the adverse impacts of 'natural' disasters and climate change are understood as failures of state protection, they are clearly incorporated into the realm of possible circumstances in which a person may establish a well-founded fear of being persecuted for a Convention reason in the context of 'natural' disasters and climate change.

However, even as the determinations by the NZIPT in *AF (Kiribati)* and *AC (Tuvalu)* reflect how a human-rights-based approach to RSD provides a clear and principled methodology, it is based on an interpretation of the refugee definition that appears to cast the temporal scope too narrowly and the personal scope too widely. The definition of being persecuted as the sustained or systemic violation of basic human rights demonstrative of a failure of state protection encourages a focus on a specific event when considering whether a person faces a 'real chance' of being persecuted, suggesting that a person will only be persecuted once the risk of exposure to harm actually accrues, hence inviting the 'imminence' parallel identified in *AF (Kiribati)*.[275] This may reflect an unduly narrow temporal scope of the notion of being persecuted.

The definition also allows a wide range of people exposed to such violations to accurately be described as being persecuted, even though such a characterisation would depart significantly from the ordinary meaning of the term, effectively replacing the phrase 'being persecuted' with that of 'being subjected to serious human rights violations.' For example, in *Budayeva*,[276] the victims of the mudslide that gave rise to the violations of the right to life under Article 2 ECHR may accurately be described as having suffered from a sustained or systemic violation of human rights demonstrative of a failure of state protection, given the systemic failures of state actors to take steps to protect the victims from the known risks associated with mudslides. However, few would contend that this clear violation of the right to life amounted to 'persecution'. Similarly, not all persons who face serious denials of economic and social rights can accurately be described as being persecuted, even if their

---

[275] Even though explanations about how the definition should be understood have at times drawn attention to the 'risk' of being persecuted being 'sustained or systemic'. See discussion in Chapter 5.
[276] *Budayeva v Russian Federation* (n 193).

predicament reveals a sustained or systemic violation of human rights. The definition casts the net too widely, and in so doing fails to draw attention to the distinctive, discrimination-based nature of the experience of being persecuted.

The leading human-rights-based definition of being persecuted therefore appears unable to effectively distinguish between people who are victims of human rights violations from people who are persecuted. Chapters 4–6 examine this definition closely, ultimately proposing a recalibrated human-rights-based interpretation. Chapter 7 then proposes a methodology for RSD that, although of general application, is considered in relation to claims for recognition of refugee status in the context of disasters and climate change.

4

Interpreting the Refugee Definition

4.1 The Unitary Character of the Refugee Definition

Before embarking upon a process of interpreting the refugee definition, it is important to emphasise its unitary character. Although substantial doctrinal and judicial work has been invested in clarifying the precise meaning of each element of the refugee definition, there is some risk of not seeing the wood for the trees. As Lord Justice Sedley observed in *Karanakaran*:

> While, for reasons considered earlier, it may well be necessary to approach the Convention questions themselves in discrete order, how they are approached and evaluated should henceforward be regarded not as an assault course on which hurdles of varying heights are encountered by the asylum seeker with the decision-maker acting as umpire, nor as a forum in which the improbable is magically endowed with the status of certainty, but as a unitary process of evaluation of evidential material of many kinds and qualities against the Convention's criteria of eligibility for asylum.[1]

Similarly, McHugh J observed in *Applicant A*:

> The phrase 'a well-founded fear of being persecuted for reasons of ... membership of a particular social group' is a compound conception. It is therefore a mistake to isolate the elements of the definition, interpret them, and then ask whether the facts of the instant case are covered by the sum of those individual interpretations. Indeed to ignore the totality of the words that define a refugee for the purposes of the Convention and the Act would be an error of law by virtue of a failure to construe the definition as a whole.[2]

In *Refugee Appeal No. 74665/03*, the case in which the New Zealand RSAA expressly adopted a 'predicament' approach to RSD, the Authority recognised the importance of applying the refugee definition as a whole, but nevertheless asserted the existence of the need to understand the meaning of the constituent elements. On that basis, it observed:

---

[1] *Karanakaran v Secretary of State for the Home Department* [2000] EWCA Civ. 11, [2000] Imm AR 271 [19], cited in Allan Mackey et al, 'A Structured Approach to the Decision Making Process in Refugee and Other International Protection Claims' (Presented at the IARLJ/JRTI/UNHCR conference: 'The Role of the Judiciary in Asylum and Other International Protection Law in Asia', Seoul, Korea, 10–11 June 2016) [5] <https://www.iarlj.org/iarlj-documents/general/IARLJ_Guidance_RSD_paper_and_chart.pdf> accessed 4 April 2019.
[2] *Applicant A v Minister for Immigration and Ethnic Affairs* [1997] HCA 4, [1997] 190 CLR 225, 256 (McHugh J).

> It is essential to ensure that one element is not inadvertently given a function or meaning which more properly belongs to another ... Thus the question whether there is a Convention ground cannot sensibly be conflated or confused with the issue of risk (the well-foundedness element). Similarly the question whether the anticipated harm can properly be described as 'being persecuted' is not an analysis which belongs in the assessment of risk.[3]

In the remaining chapters of the book, the argument is developed that features of the well-founded fear criterion and the nexus requirement have indeed been given a function and meaning that more properly belongs to the 'being persecuted' element of the refugee definition. Although the principles that have been articulated in judicial decisions and legal doctrinal analysis remain critical to the interpretation of the definition, the recalibrated interpretation that is developed in Chapters 5–6 is more faithful to the unitary character of the refugee definition, and the ordinary meaning of the individual terms in context, and in light of the object and purpose of the Convention.

4.2 The Relevance of International Human Rights Law to the Interpretation of Being Persecuted

As a point of departure, it is important to acknowledge the relevance of international human rights law (IHRL) to the interpretation of the meaning of being persecuted under the refugee definition, whilst simultaneously emphasising the distinctiveness of the legal concept of being persecuted from the legal concept of being a victim of a human rights violation. Although Hathaway's original explanation of the relevance of IHRL to this interpretative exercise draws support more from legal doctrinal sources,[4] the second edition, written together with Michelle Foster, relies more heavily on the methodology prescribed by Articles 31–33 of the Vienna Convention on the Law of Treaties (VCLT). For Hathaway and Foster, the human rights-based approach is the preferred one to adopt as a matter of treaty interpretation because of the prominent position afforded to the Universal Declaration on Human Rights in the Convention's preamble,[5] which states at the first and second perambular paragraphs:

> CONSIDERING that the Charter of the United Nations and the Universal Declaration of Human Rights ... have affirmed the principle that human beings shall enjoy fundamental rights and freedoms without discrimination,
>
> CONSIDERING that the United Nations has, on various occasions, manifested its profound concern for refugees and endeavoured to assure refugees the widest possible exercise of these fundamental rights and freedoms ...[6]

They also cite widespread judicial support for the proposition that the Convention was 'written against the background of international human rights law' and its 'clear human rights object and purpose is the background against which interpretation of individual

---

[3] *Refugee Appeal No. 74665/03* RSAA (7 July 2004) [48].
[4] See James Hathaway, *The Law of Refugee Status* (1st edn, Butterworths 1991) 108, fn 59.
[5] James Hathaway and Michelle Foster, *The Law of Refugee Status* (2nd edn, CUP 2014) 193.
[6] Convention Relating to the Status of Refugees (adopted 28 July 1951, entered into force 22 April 1954) 189 UNTS 137, first and second perambular paragraphs.

provisions must take place',[7] as well as acknowledgement by the UNHCR that the 'strong human rights language' in the preamble confirms that 'the aim of the drafters [was] to incorporate human rights values in the identification and treatment of refugees, thereby providing helpful guidance for the interpretation, in harmony with the Vienna Convention, of the provisions of the 1951 Convention'.[8] Though some have questioned this heavy reliance on the preamble to support the human rights-based approach to interpretation,[9] it is accepted here that IHRL is clearly relevant to the interpretation of the Convention. However, being persecuted is not the same as being the victim of a human rights violation.

International human rights law is primarily concerned with the obligations of states to respect, protect, and fulfil the rights guaranteed under ratified treaties and relevant customary international law of people within their jurisdiction. Specific instruments, such as the International Covenant on Economic, Social and Cultural Rights, contain general provisions (such as the prohibition on discrimination) and enumerate substantive rights, such as the right to food, the right to shelter, the right to the highest attainable standard of health, and so forth. Treaty monitoring bodies, such as the Committee on Economic, Social and Cultural Rights, the Committee on the Rights of the Child, and the Committee on the Elimination of All Forms of Discrimination against Women, supervise states parties' compliance with their treaty obligations, and also produce General Comments or Recommendations regarding the normative content of the obligations under the supervised instrument. Further insight into an instrument's normative content may be provided by judicial or quasi-judicial interpretation and application, whether at the level of an individual complaints procedure before a treaty monitoring body or regional court, or in domestic judicial settings. Academic commentary, as well as the work of Special Rapporteurs and other mandate holders, provides additional insight. These are just some of the ways in which the normative content of IHRL provisions is developed over time.

When taking a human rights-based approach to the refugee definition, it is important to emphasise that the Refugee Convention is a distinct instrument, and the phrase 'a well-founded fear of being persecuted' is not replicated within the international human rights law canon. Through the development of complementary forms of protection, similar notions, such as 'substantial grounds have been shown for believing that the person concerned ... faces a real risk of being subjected to torture or to inhuman or degrading treatment or punishment',[10] have been expressed when determining the extent of a host state's *non-refoulement* obligation. However, notwithstanding the potential policy benefits of achieving systemic integration between international refugee law and international human rights law,[11] it is important not to permit the latter to dilute the distinctiveness of the former.

---

[7] Hathaway and Foster, *Refugee Status* (n 5) 193, citing *Applicant A* (n 2) 296–99 (Kirby J) and *Pushpanathan v Canada (Minister of Citizenship and Immigration)* [1998] 1 SCR 982, 1024.

[8] Hathaway and Foster, *Refugee Status* (n 5) 194, citing UNHCR, 'The International Protection of Refugees: Interpreting Article 1 of the 1951 Convention Relating to the Status of Refugees' (2001) 20 Refug Surv Q 77, 78.

[9] See for example David Cantor, 'Defining Refugees: Persecution, Surrogacy and the Human Rights Paradigm' in Bruce Burson and David Cantor (eds), *Human Rights and the Refugee Definition: Comparative Legal Practice and Theory* (Brill Nijhoff 2016) 376.

[10] *Soering v United Kingdom* [1989] 11 EHRR 439, [91].

[11] See Hathaway and Foster, *Refugee Status* (n 5) 9.

Not only is the phrase 'a well-founded fear of being persecuted' distinctive, but the purpose of international refugee law is different from the purpose of international (and regional) human rights law. International refugee law is concerned with providing 'surrogate' protection where an individual cannot rely on her own state to fulfil its obligations to respect, protect, and fulfil her human rights. Hence, it is primarily concerned with the relationship between the individual and the international community (represented by a host state). International human rights law, in contrast, is concerned with regulating the relationship between the individual and the state. Guidance is developed to assist states to fulfil their obligations under IHRL, and judicial and quasi-judicial bodies may consider whether a state has failed to do so both in general terms and on a case-by-case basis.

Although the jurisprudence and guidelines developed in this connection are relevant for determining whether a person may accurately be described as being persecuted, this experience entails something distinct from being the victim of a human rights violation. As Hathaway and Foster note, RSD is not concerned with assessing 'liability for breach of human rights law', even if a breach may well be apparent in the majority of cases. This view has been cogently expressed by McHugh J in *Minister for Immigration and Multicultural Affairs v Respondents S152/2003*:

> In neither its ordinary nor its Convention meaning does the term 'persecution' require proof that the State has breached a duty that it owed to the applicant for refugee status. Where the State is involved in persecution, it will certainly be in breach of its duty to protect its citizens from persecution. But that is beside the point. State culpability is not an element of persecution.[12]

Rather than inviting adjudication, reference to human rights law principles can provide a tool that helps to *inform*[13] whether the harm to which a person fears being exposed attains the requisite severity threshold, since not every form of adversity or oppression can amount to 'persecution'. It warrants noting that the second edition of *The Law of Refugee Status* replaces the term 'violation' with 'denial' when defining being persecuted as 'the sustained or systemic denial of human rights demonstrative of a failure of state protection'. Although the authors do not explain this change in the text, this book embraces the term 'denial', as it helps to further shift attention to the predicament of the individual, rather than placing too much attention on whether the state has formally breached an obligation under IHRL. It is submitted here that the revised approach must be correct, as a person can accurately be described as being persecuted even in situations where the state is doing its best and yet fails to protect an individual from serious denials of human rights. This point is particularly relevant when considering the kinds of circumstances in which a person may establish a well-founded fear of being persecuted for a Convention reason in the context of disasters and climate change, as will be developed further in the remainder of the book.

---

[12] *Minister for Immigration and Multicultural Affairs v Respondents* S152/2003 [2004] 222 CLR 1 (McHugh J), cited in Jane McAdam, 'Interpretation of the 1951 Convention' in A Zimmermann (ed), *The 1951 Convention Relating to the Status of Refugees and Its 1967 Protocol: A Commentary* (OUP 2011) 89. As McAdam explains, McHugh J's comment concerned a doctrine developed in German jurisprudence that understood 'persecution' as 'State persecution'. The same reasoning, however, applies to the human rights adjudicatory approach.

[13] Hathaway and Foster, *Refugee Status* (n 5) 200 (original emphasis).

4.3 The Methodology Prescribed by the VCLT

Articles 31–33 of the VCLT establish the method to be employed for interpreting the meaning of provisions in international treaties. The product of work by the International Law Commission,[14] the VCLT reflects customary international law in this connection.[15] Consequently, even though the Refugee Convention was drafted before the entry into force of the VCLT, Articles 31–33 are still considered to guide the interpretation of that instrument.[16]

Article 31 of the VCLT provides:

Article 31, GENERAL RULE OF INTERPRETATION

1. A treaty shall be interpreted in good faith in accordance with the ordinary meaning to be given to the terms of the treaty in their context and in the light of its object and purpose.
2. The context for the purpose of the interpretation of a treaty shall comprise, in addition to the text, including its preamble and annexes:
   (a) Any agreement relating to the treaty which was made between all the parties in connexion with the conclusion of the treaty;
   (b) Any instrument which was made by one or more parties in connexion with the conclusion of the treaty and accepted by the other parties as an instrument related to the treaty.
3. There shall be taken into account, together with the context:
   (a) Any subsequent agreement between the parties regarding the interpretation of the treaty or the application of its provisions;
   (b) Any subsequent practice in the application of the treaty which establishes the agreement of the parties regarding its interpretation;
   (c) Any relevant rules of international law applicable in the relations between the parties.
4. A special meaning shall be given to a term if it is established that the parties so intended.

Of particular note, for the purposes of interpreting the meaning of being persecuted, is the stipulation in Article 31(1) that the terms of the treaty shall be interpreted 'in accordance with the ordinary meaning to be given to the terms ... in their context and in light of its object and purpose'. It is recognised that there is no hierarchy between these three elements. Rather, reference must be made to the context and to the object and purpose of the treaty when discerning the ordinary meaning of the terms.[17] Importantly, Article 31(2) provides that the context includes the preamble to the treaty.

---

[14] See Draft Articles on the Law of Treaties in Yearbook of the International Law Commission 1966 vol. II (1967) (A/CN.4/SER. A/1966/Add. 1).

[15] See Oliver Dörr and Kirsten Schmalenbach, 'Article 31: General Rule of Interpretation' in Oliver Dörr and Kirsten Schmalenbach (eds), *Vienna Convention on the Law of Treaties: A Commentary* (Springer 2012) 523, citing ICJ Arbitral Award of 31 July 1989 (Judgment) [1991] ICJ Rep 53, [48].

[16] See for example *Golder v United Kingdom* [1979] 1 EHRR 524, [29] in Malagosia Fitzmaurice, 'Interpretation of Human Rights Treaties' in Dinah Shelton (ed), *The Oxford Handbook of International Human Rights Law* (OUP 2013) 759.

[17] This approach is referred to as the 'crucible approach' and has been applied in practice by the ICJ in, for example, *Avena and Other Mexican Nationals (Mexico v United States)* [2004] ICJ Rep 12, from [83], cited in Fitzmaurice, 'Interpretation of Human Rights Treaties' (n 16) 746–47.

Multiple references have been made to the 'object and purpose' of the Refugee Convention throughout this book,[18] with a general recognition that the Convention exists to provide a kind of 'surrogate protection', and is inspired by 'general underlying themes of the defence of human rights and anti-discrimination'[19] reflected in judicial opinion,[20] as well as academic commentary.[21] The approach to interpretation adopted here does not diverge from that consensus.[22]

Reference at Article 31(3)(c) to 'any relevant rules of international law applicable in the relations between the parties' is also significant when a human rights-based approach is adopted as, together with the human rights references in the preamble, and the 'broad human rights purpose' of the Convention as a whole, the canon of IHRL that is applicable in the relations between the parties falls expressly to be considered. However, as Cantor notes,[23] it is not strictly possible to identify those instruments that are 'applicable in the relations between the parties', as none of the human rights law instruments is universally ratified. As Chapter 6 develops more fully, the non-discrimination norm, as potentially a general norm of customary international law, or simply because it finds expression in the UN Charter as well as in the UDHR and across the human rights treaties, may be the one norm that can claim such a wide applicability.

Article 32 concerns recourse to supplementary means of interpretation, including preparatory works where, after following the procedure set out at Article 31, the meaning of the terms '(a) Leaves the meaning ambiguous or obscure; or (b) Leads to a result which is manifestly absurd or unreasonable'. The interpretation of being persecuted that is performed over Chapters 5–6 suggests that the meaning of that term, and indeed of the compound notion of being persecuted for a Convention reason, is neither ambiguous nor obscure, neither does it lead to a result that is manifestly absurd or unreasonable. Of course, the existence of multiple interpretations of the definition in refugee law doctrine and jurisprudence would suggest otherwise, and reference to the *travaux* for insight into the meaning of the notion of being persecuted may well be appropriate. However, it warrants noting that Einarsen, in consulting the preparatory works for his contribution to the Zimmerman volume, found that:

> [a] reasonable inference is that the drafters in 1950 reckoned 'persecution' to be a well-known concept, taking into account the vast and basically European experience with persecutions, refugees, and different mass flight situations over the previous 30 years.

---

[18] See for example Section 1.2 and Section 3.2.5.
[19] *Canada (Attorney General) v Ward* [1993] 2 SCR 689.
[20] See for example *R v Secretary of State for the Home Department, ex parte Sivakumaran and Conjoined Appeals* (*UN High Commissioner for Refugees Intervening*) [1987] UKHL 1, [1988] AC 958); *Ward* (n 19); *Applicant A* (n 2) 248 (Dawson J); *Horvath v Secretary of State for the Home Department* [2000] UKHL 37, [2001] AC 489 (Lord Hope); *BG (Fiji)* [2012] NZIPT 800081.
[21] Hathaway and Foster, *Refugee Status* (n 5) 193; Guy S. Goodwin-Gill and Jane McAdam, *The Refugee in International Law* (3rd edn, OUP 2007) 7–8: 'For the 1951 Convention relating to the Status of Refugees, this means interpretation by reference to the object and purpose of extending the protection of the international community to refugees, and assuring to "refugees the widest possible exercise of ... fundamental rights and freedoms"'.
[22] The additional object and purpose relating to the development of an effective international system for addressing the refugee phenomenon is discussed in McAdam, 'Interpretation of the 1951 Convention' (n 12) 91–93.
[23] Cantor, 'Defining Refugees' (n 9) 376–77.

One gets the impression that the drafters simply thought they knew who was a 'refugee', and that the application of a general definition would not be so difficult in practice.[24]

Finally, Article 33 of the VCLT addresses interpretation of treaties authenticated in two or more languages. This provision is relevant because the Refugee Convention is authentic in both its English and French versions. However, as the French *'craignant avec raison d'être persécutée'* differs only by drawing attention away from any notion of a subjective 'trepidation' element,[25] there is no reason to examine closely the provisions of Article 33.

To present Articles 31–33 in the manner in which they appear within the VCLT suggests that treaty interpretation is a practice that, if conducted in accordance with this methodology, will yield a clear and principled interpretation. As it turns out, authorities on treaty interpretation recognise that arriving at 'one right answer' is unlikely to result from the faithful application of Articles 31–33.[26] In the context of the interpretation of international refugee law, substantial academic and judicial authority supports the Hathaway definition of being persecuted as the sustained or systemic violation of human rights demonstrative of a failure of state protection.[27] The weight of authority would thus suggest that Hathaway has arrived at, if not the 'correct' interpretation, then at least a clear and principled interpretation that is faithful to the principles of interpretation set out at Articles 31–33 of the VCLT. However, as Chapters 5–6 argue, the Hathaway definition pays insufficient attention to the ordinary meaning of the term, and thus unduly narrows the temporal scope of the predicament of being persecuted, whilst simultaneously providing for an unduly broad personal scope that would recognise anyone exposed to a sustained or systemic violation of human rights as being persecuted.

---

[24] Terje Einarsen, 'Drafting History of the 1951 Convention and the 1967 Protocol' in A Zimmermann (ed), *The 1951 Convention Relating to the Status of Refugees and Its 1967 Protocol: A Commentary* (OUP 2011) 56–57.

[25] See for example James Hathaway and William Hicks, 'Is There a Subjective Element in the Refugee Convention's Requirement of "Well-Founded Fear"?' (2005) 26 Mich J Intl L 505, 538.

[26] Ulf Linderfalk, *On the Interpretation of Treaties: The Modern International Law as Expressed in the 1969 Vienna Convention on the Law of Treaties* (Springer 2010), 4–5.

[27] See references in Hathaway and Foster, *Refugee Status* (n 5) 196–97.

5

The Temporal Scope of Being Persecuted

5.1 Being Persecuted from the Perspective of the 'Event Paradigm'

In *Acosta*, the Board of Immigration Appeals defined 'persecution' as 'the infliction of suffering or harm in order to punish an individual for possessing a particular belief or characteristic the persecutor seeks to overcome'.[1]

As Chapter 3 established, international refugee law doctrine (if not domestic jurisdictions) has largely rejected the requirement of establishing the malign intention of the actor of persecution.[2] However, despite recognition of the doctrinal salience of the predicament approach, much focus remains on the nature of the harm to which an individual is exposed.

That the quality of the harm dominates doctrinal understanding of what being persecuted means is nowhere more evident than in the approach adopted by the International Association of Refugee Law Judges (IARLJ) (European Chapter), whose Judicial Analysis of the recast Qualification Directive[3] explains that '[t]he Refugee Convention provides no definition of the term "being persecuted" but one is provided in EU law through Article 9(1) QD (recast)'.[4]

'Being persecuted' is translated in the language of the Qualification Directive, and in the minds of leading international refugee law judges, into the acts that cause the qualifying harm. In fact, the Qualification Directive does not provide a definition of 'being persecuted', because, as highlighted below, Article 9 provides a definition of *acts* of persecution. These are two different, albeit interrelated, notions. The IARLJ, and the state of the art in international refugee law doctrine, tends to elide the two. This elision reflects an 'event paradigm' in international refugee law. Such an approach narrows the temporal scope of the refugee definition to the moment the harm is experienced, and thus detracts attention from the wider social context in which the risk of exposure to such acts arises.

---

[1] *Matter of Acosta, A-24159781*, United States Board of Immigration Appeals, 1 March 1985, cited in Guy S Goodwin-Gill and Jane McAdam, *The Refugee in International Law* (3rd edn, OUP 2007) 91.
[2] See discussion in Section 3.2.
[3] Council of the European Union, Directive 2011/95/EU of the European Parliament and of the Council of 13 December 2011 on Standards for the Qualification of Third-Country Nationals or Stateless Persons as Beneficiaries of International Protection, for a Uniform Status for Refugees or for Persons Eligible for Subsidiary Protection, and for the Content of the Protection Granted (Recast), 20 December 2011, OJ L. 337/9-337/26; 20.12.2011 (Qualification Directive).
[4] IARLJ-Europe, 'Qualification for International Protection (Directive 2011/95/EU): A Judicial Analysis' (EASO 2016) 27 <https://www.easo.europa.eu/sites/default/files/QIP%20-%20JA.pdf> accessed 15 April 2019.

The elision of the predicament of being persecuted with acts of persecution highlighted above is supported by leading doctrinal authorities. Grahl-Madsen, writing in 1966, considered what he termed 'liberal' and 'restrictive' schools of thought concerning the meaning of 'persecution', both of which reflect a focus on the nature of the harm, as distinct from the predicament of the claimant. Paul Weis, for the 'liberal' school, considered that 'other measures [than a threat to life or freedom] in disregard of human dignity may also constitute persecution',[5] whilst Jacques Vernant, also of the 'liberal' school, considered that 'severe measures and sanctions of an arbitrary nature, incompatible with the principles set forth in the Universal Declaration of Human Rights', were equivalent to 'persecution'.[6] The contrasting 'restrictive' school, represented by Karl Friedrich Zink, sees persecution as meaning 'only deprivation of life or of physical freedom'.[7]

In contemporary doctrine, Goodwin-Gill and McAdam, for example, first note that '"[p]ersecution" is not defined in the 1951 Convention or in any other international instrument',[8] before considering the range of acts that may amount to 'persecution', including torture as well as 'acts covered by the prohibition on cruel, inhuman or degrading treatment or punishment, or punishment, or repeated punishment for breach of the law, which is out of proportion to the offence'.[9] Persecution happens the moment a person experiences the qualifying harm. This event-based approach to 'persecution' is echoed by Zimmermann and Mahler:

> To determine whether a severe violation of human rights amounting to persecution has been taking place, a complex bundle of factors ... has to be taken into consideration. Possible factors are the intensity of the acts and their duration; the danger of or the actual recurrence of such acts; whether the acts occur only in individual cases or as parts of a larger campaign of systematic human rights violations; and finally the effect of such acts on the health, family life, or participation in political life of the person concerned.[10]

Similarly, Hathaway explains there that 'the refugee claimant must apprehend a form of harm which can be characterized as "persecution"'.[11] However, the Hathaway definition actually signals a wider temporal scope than these decidedly event-based approaches, owing to the inclusion of the requirement that a denial of human rights be 'sustained or systemic'. In the first edition of *The Law of Refugee Status*, Hathaway draws on the characterisation by the Canadian Immigration Appeals Board of persecution as 'harassment that is so constant and unrelenting that the victims feel deprived of all hope of recourse, short of flight, from government by oppression',[12] to argue that '[t]he equation of persecution with harassment highlights the need to show a sustained or systemic risk, rather than just an isolated incident of harm'.[13]

---

[5] Paul Weis, UN Doc. HCR/INF/49, 22, cited in Atle Grahl-Madsen, *The Status of Refugees in International Law* (AW Sijthoff's Uitgeversmaatschappij NV 1966), 193.
[6] Jacques Vernant, *The Refugee in the Post-War World* (George Allen & Unwin 1953) 8, cited in Grahl-Madsen, *The Status of Refugees in International Law* (n 5) 193.
[7] Ibid   [8] Goodwin-Gill and McAdam, *The Refugee in International Law* (n 1) 90.   [9] Ibid 90–91.
[10] Andreas Zimmermann and Claudia Mahler, 'Article 1A, para. 2 (Definition of the Term "Refugee"/ Définition du Terme "Réfugié")' in A Zimmermann (ed), *The 1951 Convention Relating to the Status of Refugees and Its 1967 Protocol: A Commentary* (OUP 2011) 348.
[11] James Hathaway, *The Law of Refugee Status* (Butterworths 1991) 99.
[12] *Gladys Maribel Hernandez*, Immigration Appeal Board Decision M81-1212, January 6, 1983, cited in Hathaway, *Refugee Status* (n 11) 102.
[13] Ibid

Recognising that the 'sustained or systemic' element of the definition of being persecuted has not always been well understood, Hathaway and Foster clarify in the second edition of *The Law of Refugee Status* that a single act is capable of attaining the persecution threshold.[14] It is not the harm itself that must be sustained or systemic, but rather the *risk of harm*:

> [T]he human rights approach is ideally suited to the task of identifying serious harm for purposes of knowing whether an individual faces *the risk* of 'being persecuted'. So long as *the risk* of denial of a broadly accepted international human right is sustained – in the sense that, as a practical matter, it is ongoing; or systemic – in the sense that *the risk* is endemic to the political or social system – it can reasonably be said that there is *a risk* of 'being persecuted' of the kind that may engage Convention obligations.[15]

Understood thus, the sustained or systemic element of the definition appears superfluous to what appears in practice as the bifurcated approach reflected in the 'serious harm + failure of state protection' formula.[16] A definition of being persecuted as the sustained or systemic *risk* of a serious denial of human rights demonstrative of a failure of state protection would have been a more accurate characterisation of the predicament, but this is not what has been advanced.

Hugo Storey, in his attempt to formulate a systematic definition of 'persecution', recognises the significant contribution of Hathaway's approach, but, taking the sustained or systemic element of the definition at face value rather than according it the function outside of the being persecuted definition that Hathaway and Foster give it in their text, would do away with it altogether:

> [T]he requirement that violations of human rights must always be 'sustained or systemic' and must involve some level of persistency or repetition would, however, entail treating single acts of serious harm as non-persecutory, even if they took the form of murder or torture. That is plainly too restrictive.[17]

Drawing on Lambert,[18] Storey provides support for this position with reference to Australian[19] and UK[20] jurisprudence, as well as institutional[21] and academic[22] commentary

---

[14] James Hathaway and Michelle Foster, *The Law of Refugee Status* (2nd edn, CUP 2014) 195 fn 78.
[15] Ibid 185 (emphasis added).
[16] *Islam v Secretary of State for the Home Department Immigration Appeal Tribunal and Another, ex parte Shah, R v* [1999] UKHL 20, [1999] AC 629, 653 (Lord Hoffman), cited in Hathaway and Foster, *Refugee Status* (n 14) 185: 'As this well-accepted formulation makes clear, the phrase "being persecuted" comprises two essential elements, succinctly expressed by Lord Hoffman as "persecution = serious harm + failure of state protection"'.
[17] Hugo Storey, 'Persecution: Towards a Working Definition' in Vincent Chetail and Céline Bauloz (eds), *Research Handbook on International Law and Migration* (Edward Elgar 2014) 472–73.
[18] Hélène Lambert, 'The Conceptualisation of "Persecution" by the House of Lords: Horvath vs Secretary of State' (2001) 13 IJRL 16, 28.
[19] *Minister for Immigration v Haji Ibrahim* [2000] HCA 55, [2000] 204 CLR 1.
[20] *Doymus v Secretary of State for the Home Department* (IAT, HX/80112/99, 19 July 2000).
[21] UNHCR, *Handbook and Guidelines on Procedures and Criteria for Determining Refugee Status under the 1951 Convention and the 1967 Protocol Relating to the Status of Refugees* (Reissued December 2011) [51] and [53] <http://www.unhcr.org/publications/legal/3d58e13b4/handbook-procedures-criteria-determining-refugee-status-under-1951-convention.html> accessed 4 April 2019.
[22] Lambert, 'The Conceptualisation of "Persecution"' (n 18) 23 fn 32, who relies on Deborah Anker, *Law of Asylum in the United States* (3rd edn, Refugee Law Centre 1999) and Guy S Goodwin-Gill, *The Refugee in International Law* (Clarendon 1996).

rejecting or not adopting any requirement for 'persistency' or systematicity in the interpretation of being persecuted. Lambert explains how the IAT in *Doymus* found that:

> the requirement of 'persistency' runs counter to the logic of Hathaway's analysis of persecution by reference to a hierarchy of rights, as well as common sense; a single incident affecting an individual's absolute right should suffice. It reached the convincing conclusion that, in view of the survey of the jurisprudence, 'persistency is a usual but not a universal criterion of persecution'.[23]

Where the 'sustained or systemic' feature of 'persecution' appears relevant, according to Storey, is when considering 'claims to refugee status based, not on individual risk characteristics, but rather membership of a class or category of persons'. 'In such cases', he continues, 'it makes sense that persons should be required to show that the scale and frequency of problems facing similarly situated persons are at a level where it is possible to say that the generality of such persons face harm'.[24]

This clarification mirrors Hathaway and Foster's explanation that the sustained or systemic element of their definition refers to the risk of harm, not the harm itself. However, Storey rightly finds difficulty with the definition when read at face value.[25] In this connection, Storey celebrates the drafting of Article 9(1) of the Qualification Directive, which he considers the 'one major development' since 'Hathaway gifted us his human rights theory':[26]

> It can be seen that article 9(1) offers a shorthand definition that is essentially Hathaway's save for a salutary modification that avoids making persistency a necessary condition: note the italicised words from article 9(1)(a): '*by their nature or* repetition ...'.[27]

It is in this way that a person who faces being shot and killed can establish a well-founded fear of being persecuted. On this model, a person is accurately described as persecuted when she is shot and killed. The element of risk that exists within Hathaway and Foster's explanation of the 'sustained and systemic' formulation is lost with the elision of being persecuted with 'acts of persecution' under Article 9(1). Risk is not a feature of being persecuted, but rather falls to be considered elsewhere, presumably, as developed in Section 5.2, within the well-founded fear criterion.

Although 'persecution' is understood as an act that is inflicted on a person, Storey does recognise that the temporal scope of the refugee definition as a whole is wider than the instant of persecution. He notes that:

---

[23] Lambert, 'The Conceptualisation of "Persecution"' (n 18) 22–23.
[24] Storey, 'Persecution: Towards a Working Definition' (n 17) 473.
[25] See further Hugo Storey, 'The Law of Refugee Status, 2nd edition: Paradigm Lost?' (2015) 27 IJRL 348, 356–57 where Storey opines: 'One has the feeling here that the authors are simply not prepared to acknowledge the problem identified by a number of courts and tribunals with the unqualified use of adjectives such as "sustained" or "systemic". For authors who in many places demand analytical rigour of others, it is not good enough that they simply restate the formulation and leave it to a footnote to acknowledge that it is meant by them to be read as a defeasible proposition'.
[26] Hugo Storey, 'What Constitutes Persecution? Towards a Working Definition' (2014) 26 IJRL 272, 279.
[27] Ibid 280.

Reference to the forward-looking nature of the test is, however, an important reminder of the need to understand current fear as on-going fear and not just confined to the immediate present. If there is no well-founded fear today but will be tomorrow then it is impossible to say the claimant will be protected against persecution.[28]

Yet again, risk is not a feature of the experience of being persecuted, but rather a factor to consider in determining whether a person may be persecuted in the future.

In contrast to the perspective advanced by Storey above, the NZIPT expresses an understanding of the Hathaway definition in which the sustained or systemic element remains an essential component of being persecuted:

> 'Systemic' (not, as sometimes stated, systematic) identifies that 'being persecuted' arises because of an anticipated failure of the legal and other protection-relevant systems in the claimant's country of origin. Finally, 'sustained' can be seen to serve two functions. It references the enduring nature of the claimant's predicament arising from the failure of state protection in the country of origin. It also reminds decision-makers that persecutory harm can, but not must, encompass multiple and ongoing violations of rights.[29]

On this approach, which is more faithful to the explanation provided by Hathaway and Foster, being persecuted is understood with reference to the 'enduring nature of the claimant's predicament'. However, this nuance is readily lost under the repeated doctrinal and judicial elision of being persecuted with acts of persecution, evident indeed even in the NZIPT framing.

As the following section argues, the 'event paradigm' described above is reinforced by the interpretation of the well-founded fear element as entailing a risk assessment.

5.2 The 'Well-Founded Fear' Element as Risk Assessment

The 'event paradigm' that dominates interpretations of the being persecuted element of the refugee definition is reinforced by the idea that the well-founded fear element requires a risk assessment. When being persecuted is understood as an act that might happen, then determining whether a person's fear of being persecuted is 'well-founded' logically entails asking about the likelihood of a risk accruing.

Indeed, by replacing the formulation 'well-founded fear' with interpretations such as 'reasonable likelihood'[30] or 'real chance',[31] doctrine narrows the refugee experience to isolated instances of serious harm, by focusing attention on the likelihood of a specific event occurring, rather than directing decision-makers to take a holistic approach to the claimant's predicament. The parallels with the 'real risk of serious harm' and similar formulations in complementary protection regimes reflect this preoccupation with the moment of harm.[32]

---

[28] Storey, 'Persecution: Towards a Working Definition' (n 17) 491.
[29] *DS (Iran)* [2016] NZIPT 800788, [126].
[30] *R v Secretary of State for the Home Department, ex parte Sivakumaran and Conjoined Appeals (UN High Commissioner for Refugees Intervening)* [1987] UKHL 1, [1988] AC 958.
[31] *Chan v Minister for Immigration and Ethnic Affairs* [1989] HCA 62, [1989] 169 CLR 379.
[32] See HRC, General Comment No. 31/80: The Nature of the General Legal Obligation Imposed on States Parties to the Covenant (26 May 2004) CCPR/C/21/Rev.1/Add.13 [12] where the notion of a 'real risk of irreparable harm' is used; Council of the European Union, Directive 2011/95/EU (n 3) Article 15; *Cruz*

Early case law and doctrine helps to illustrate the event paradigm through its transformation of the well-founded fear criterion into a 'reasonable likelihood' or 'real chance' test.

Without repeating the entirety of the argument, it is helpful to begin with reference to Cox's 1984 article[33] tracing the development of the well-founded fear criterion and its firm roots in earlier formulations including 'reasonable grounds' and 'good reasons'. Cox begins with the Constitution of the International Refugee Organization (IRO), which was adopted by the UN General Assembly on 15 December 1946.[34] The refugee definition at Annex I, Section A of the Constitution[35] does not include the term 'well-founded fear', but is rather entirely constructed on the basis of past persecution. Reflecting state practice in relation to earlier refugee flows,[36] refugee status under the IRO definition was based on categories, such as victims of Nazi or fascist regimes. However, a more general category of 'persons who were considered refugees before the outbreak of the second world war, for reasons of race, religion, nationality or political opinion' was also included in the definition.

Consideration of the conditions that would await such refugees on return arose only in relation to whether an individual had 'valid objections' to returning. Under Section C – Conditions under which 'refugees' and 'displaced persons' will become the concern of the organization', 'valid objections to returning' include:

> persecution, or fear, *based on reasonable grounds* of persecution because of race, religion, nationality or political opinions.[37]

Cox goes on to describe the process by which the notion of 'fear, based on reasonable grounds', was transformed over successive proposals into a 'good reasons' criterion, which was in turn replaced with the notion of a 'well-founded fear' criterion.[38] The idea that a refugee must be able to provide 'a plausible account' also informed how the drafters understood the 'well-founded fear' criterion, as Cox explains:

Varas and Others v *Sweden* (1991) 14 EHRR 1, [69] where the notion of a real risk of being subjected to torture or to inhuman or degrading treatment or punishment is used.

[33] Theodore N Cox, '"Well-Founded Fear of Being Persecuted": The Sources and Application of a Criterion of Refugee Status' (1984) 10 Brook J Int'l L 333.

[34] United Nations, 'Constitution of the International Refugee Organization' (15 December 1946) 18 UNTS 3.

[35] 1. Subject to the provisions of sections C and D and of Part II of this Annex, the term 'refugee' applies to a person who has left, or who is outside of, his country of nationality or of former habitual residence, and who, whether or not he had retained his nationality, belongs to one of the following categories:

  (a) victims of the Nazi or fascist regimes or of regimes which took part on their side in the second world war, or of the quisling or similar regimes which assisted them against the United Nations, whether enjoying international status as refugees or not;
  (b) Spanish Republicans and other victims of the Falangist regime in Spain, whether enjoying international status as refugees or not;
  (c) persons who were considered refugees before the outbreak of the second world war, for reasons of race, religion, nationality or political opinion.

[36] See for example League of Nations, 'Convention Relating to the International Status of Refugees', 28 October 1933, League of Nations, Treaty Series Vol. CLIX No. 3663, which concerned the international status of Armenians, Russians, and 'assimilated refugees'. 'Assimilated refugees' included Assyrians, Assyro-Chaldeans, Syrians, Kurds, and 'a small number of Turks' – see Gilbert Jaeger, 'On the History of the International Protection of Refugees' (2001) 83 Int Rev Red Cross 727, 730.

[37] United Nations, 'Constitution of the International Refugee Organization' (n 34) (emphasis added).

[38] Ibid 344–50.

Two official comments were made by the Ad Hoc Committee concerning the meaning of 'well-founded fear'. The first, in a Draft Report of February 15, 1950, states that 'well-founded fear' means that a person can give a 'plausible account' of why he fears persecution. The comment contained in the Final Report of the Ad Hoc Committee, however, provides that 'well-founded fear' means that a person can show 'good reason' why he fears persecution.[39]

A 'good reasons' test, as with earlier-proposed 'reasonable grounds' and 'plausible account' approaches, all clearly reflect that the drafters intended the well-founded fear criterion to establish a standard of proof by which a person's reasons for resisting return to her country of origin or habitual residence could be evaluated. This review provides strong support for the contention that the well-founded fear criterion reflects a standard of proof, not a risk assessment.

However, what appears relatively clear quickly became less so, with early commentary by Grahl-Madsen introducing some uncertainty into the meaning of the 'well-founded' element of the refugee definition. He appears somewhat undecided on the appropriate interpretation, identifying its function as both a standard of proof[40] and a risk assessment. In relation to this first function, agreeing with the Bayerisches Verwaltungsgericht Ansbach in *Case 3719 II/58* (25 March 1959), he recognises that the criterion is fulfilled 'when a reasonable person would draw the conclusion from the external facts that he would be subject to persecution in his home country'.[41] Thus, in Grahl-Madsen's initial view, the 'reasonable person' test applies to the well-founded element. The standard of proof function of the well-founded element is further demonstrated by Grahl-Madsen's reference to the Ad Hoc Committee's report, which states that a person's fear of being persecuted will be well-founded when 'a person has actually been a victim of persecution or can show good reasons why he fears persecution'.[42]

However, within the same discussion of *Case 3719 II/58*, Grahl-Madsen introduces the notion of 'likelihood' of persecution, expressing his view that the Court in *Case 3719 II/58* 'appropriately links the concept of "well-founded fear" with "external facts" and the likelihood of persecution'.[43] Indeed, referring to the Ad Hoc Committee's report, he reads the reference to 'good reasons' as suggesting 'any indication of the *likelihood* of future persecution'.[44]

Zimmermann and Mahler also see the well-founded fear criterion as entailing a risk assessment,[45] and Goodwin-Gill and McAdam confirm:

> For the heart of the question is whether that 'subjective' fear is well-founded; whether there are sufficient facts to permit the finding that this applicant, in his or her particular circumstances, faces a serious possibility of persecution.[46]

---

[39] Ibid 349.
[40] UNHCR explains, in a document concerning the standard of proof in RSD, that '[t]he terms "burden of proof" and "standard of proof" are legal terms used in the context of the law of evidence in common law countries. In those common law countries which have sophisticated systems for adjudicating asylum claims, legal arguments may revolve around whether the applicant has met the requisite "standard" for showing that he/she is a refugee ...'. See UNHCR, 'Note on Burden and Standard of Proof in Refugee Claims' (16 December 1998) 1.
[41] Grahl-Madsen, *The Status of Refugees in International Law* (n 5) 174.
[42] Ibid 176, citing UN Doc E/11018 (E/AC-32/5, 39).   [43] Ibid 174.   [44] Ibid 179 (emphasis added).
[45] Zimmermann and Mahler, 'Definition of the Term "Refugee"' (n 10) 341–42.
[46] Goodwin-Gill and McAdam, *The Refugee in International Law* (n 1) 64.

Thus, doctrine sees RSD as entailing an exercise in determining the likelihood of something called 'persecution' happening to a person on return to her country of nationality or former habitual residence. On this approach, a person's fear is of an event that might happen.

This elision of standards of proof and risk assessment is repeated in the jurisprudence. The US Supreme Court Judgment in *Cardoza-Fonseca*[47] cements the elision of a 'well-founded fear' with 'a real risk'. In this case, the government had argued that a domestic statutory standard (which was found in a provision relating to the circumstances in which the Attorney General may be required to withhold deportation) of 'more likely than not' was applicable in the context of RSD. At first instance, the Immigration Judge dismissed the appeal, finding that the claimant had not established 'a clear probability of persecution'.[48] The Court of Appeal disagreed with this standard, and it fell to the Supreme Court to clarify the distinction between this 'clear probability' test and the 'well-founded fear' standard. The Court held that there was a clear distinction to be made between that language and the phrase 'well-founded fear' found within the domestic provision relating to the grant of asylum to a refugee:

> That the fear must be 'well-founded' does not alter the obvious focus on the individual's subjective beliefs, nor does it transform the standard into a 'more likely than not' one. One can certainly have a well-founded fear of *an event* happening when there is less than a 50% chance of the occurrence taking place.[49]

The 'event paradigm' is clearly in operation here, making it appear logical that what falls to be considered under the well-founded fear element is the risk of such an event occurring.[50] Of course, standards of proof such as the balance of probabilities (reflected in the phrase 'more likely than not') are not actuarial in nature, but rather are applied to historical realities in order to determine, to a particular degree of certainty, whether such purported historical realities can be treated as established fact. This function of the standard of proof is very different to the question about the likelihood of an event occurring in the future.

The approach taken in *Cardoza-Fonseca* was expressly followed by the UK House of Lords in *Sivakumaran*.[51] The question before the House of Lords asked whether the 'well-founded' element of the 'well-founded fear' criterion permitted a claimant's subjective fear to be objectively justified solely with reference to the facts known to the claimant or instead required reference to the full range of facts known to the Secretary of State, including those unknown to the claimant. Their Lordships unanimously agreed that the assessment entailed consideration of all facts, not only those known to the claimant, but the nature of the question inevitably affected how the case was argued and determined. Following the same dual track paved by Grahl-Madsen, counsel for the High Commissioner for Refugees had argued that use of terms such as 'good reason' and 'reasonable grounds' in the preparatory works suggested that the well-founded fear criterion requires the claimant to establish 'on the basis of the objective facts as ascertained by the Secretary of State, a reasonable man, in the position of the applicant, would think that there was a risk'.[52]

Considering that expressions 'such as "good reason" or "reasonable grounds" for fear of persecution, present questions of interpretation similar to those which arise as to the

---

[47] *INS v Cardoza-Fonseca*, 480 US 421 [1987].   [48] Ibid 425.   [49] Ibid 431 (emphasis added).
[50] The approach to the well-founded fear criterion that emphasises the subjective element of that criterion is also clearly adopted here. See discussion in Section 3.2.
[51] *Sivakumaran* (n 30).   [52] *Sivakumaran* (n 30) 998.

meaning of "well-founded fear" in the Convention itself,[53] Lord Keith of Kinkel preferred the test developed in Lord Diplock's interpretation of section 4(1)(c) of the Fugitive Offenders Act 1967[54] in *R v Governor of Pentonville Prison, ex parte Fernandez*:[55]

> My Lords, bearing in mind the relative gravity of the consequences of the court's expectation being falsified either in one way or in the other, I do not think that the test of the applicability of para (c) is that the court must be satisfied that it is more likely than not that the fugitive will be detained or restricted if he is returned. A lesser degree of likelihood is, in my view, sufficient and I would not quarrel with the way in which the test was stated by the magistrate or with the alternative way in which it was expressed by the Divisional Court. 'A reasonable chance', 'substantial grounds for thinking', 'a serious possibility' – I see no significant difference between these various ways of describing the degree of likelihood of the detention or restriction of the fugitive on his return which justifies the court in giving effect to the provisions of s 4(1)(c).[56]

Of course, as this chapter argues, there is a significant difference between tests such as 'substantial grounds for thinking', which is a standard of proof, and 'a reasonable chance' or 'a serious possibility', which are predictions about the likelihood of an event occurring in the future.

Lord Keith then applies this reasoning to the 'well-founded fear' criterion of Article 1A (2):

> I consider that this passage appropriately expresses the degree of likelihood to be satisfied in order that a fear of persecution may be well founded.[57]

Clearly Lord Keith has adopted the 'event-based' approach to the being persecuted requirement, and therefore considers that the well-founded fear element entails an assessment of the risk, or likelihood 'that persecution might indeed take place'.[58]

The interpretation of the well-founded fear criterion as requiring 'that there has to be demonstrated a *reasonable degree of likelihood* of ... persecution for a Convention reason' is then formulated by Lord Goff, appreciating the clarification provided by Lord Keith.[59]

Chan[60] is the Australian authority for the 'real chance' approach to the well-founded fear criterion. In this case, the High Court of Australia considered the predicament of a Chinese national who had been deemed by the authorities to be 'against the ideas or policies of the State'.[61] The delegate for the Minister of Immigration and Ethnic Affairs had concluded that the claimant did not have a well-founded fear of persecution because he had expressed a wish to be returned to China, rather than Macau or Hong Kong, if he were indeed to be removed from Australia. Following *Cardoza-Fonseca* and *Sivakumaran*, the High Court unanimously endorsed a 'real chance' standard, which was articulated by McHugh J (giving the lead judgment) thus:

> As the U.S. Supreme Court pointed out in Cardoza-Fonseca an applicant for refugee status may have a well-founded fear of persecution even though there is only a 10 per

---

[53] Ibid 995.
[54] Quoted by Lord Keith as providing that a person shall not be returned under the Act if it appears 'that he might, if returned, be prejudiced at his trial or punished, detained or restricted in his personal liberty by reason of his race, religion, nationality or political opinions', *Sivakumaran* (n 30) 994.
[55] *R v Governor of Pentonville Prison, ex parte Fernandez* [1971] 2 All ER 691, [1971] 1 WLR 987.
[56] *Sivakumaran* (n 30) 994. [57] Ibid 995. [58] Ibid 993. [59] Ibid 1000 (emphasis added).
[60] *Chan* (n 31). [61] Ibid 395.

cent chance that he will be shot, tortured or otherwise persecuted. Obviously, a farfetched possibility of persecution must be excluded. But if there is a real chance that the applicant will be persecuted, his fear should be characterised as 'well-founded' for the purpose of the Convention and Protocol.[62]

Again, the focus on acts of persecution, such as being shot, being tortured or 'otherwise persecuted', supports the approach under which it makes sense to speak of a risk of persecution, as distinct from having a well-founded fear of being persecuted. To recall earlier arguments, such an approach as reflected in McHugh J's opinion reveals little difference between the experience of being persecuted and the experience of suffering a serious denial of human rights.

Persecution, on this approach, is synonymous with being shot, and the likelihood of that occurring is what the well-founded fear criterion is designed to elicit. As a consequence, the event paradigm is reinforced and the focus for RSD is narrowed to an attempt to divine the likelihood of 'persecution' happening. The NZIPT's willingness to recognise similarity between the 'real chance' test under its interpretation of the refugee definition and the 'imminence' test under some forms of complementary human rights-based standards is understandable in light of this observation.[63]

Significantly, leading scholars of international refugee law have now unequivocally rejected the notion that imminence has any role to play in refugee status determination. Drawing on an extensive review of doctrine and jurisprudence, Adrienne Anderson, Michelle Foster, Hélène Lambert, and Jane McAdam argue that 'the notion of imminence should [not] be used to limit international protection in any way'.[64] They continue:

> Our critique is that because the role and significance of time is simply not acknowledged by most decision-makers, there is the potential for imminence to creep erroneously into substantive assessments, as has been seen in the examination of jurisprudence above.[65]

However, although the authors emphasise the possibility of a person satisfying the well-founded fear test even when the harm is only likely to eventuate in the foreseeable future, they continue to understand 'persecution' from within the event paradigm, arguing that 'the point in time at which an event may occur is a contextual factor in assessing risk because it can affect how likely something is to occur'.[66]

RSD under the event paradigm clearly separates risk from the experience of being persecuted. The experience of being persecuted is temporally constrained to the moment a serious denial of human rights takes place, even if the horizon for considering when that event is likely to happen is potentially distant.[67] This understanding of being persecuted as an event and the well-founded fear criterion as requiring an assessment of the risk of such

---

[62] Ibid 429 (McHugh J).   [63] See NZIPT on risk in *AF (Kiribati)* [2013] NZIPT 800413, [89].
[64] Adrienne Anderson and others, 'Imminence in Refugee and Human Rights Law: A Misplaced Notion for International Protection' (2019) 68 ICLQ 111, 139.
[65] Ibid 139–40.   [66] Ibid 139.
[67] A slightly wider temporal horizon has been envisaged by the Federal Court of Australia, although still within the event paradigm: 'It is true that a finding that there is no real chance that an applicant will suffer persecution for some time after his or her return to the country of nationality may make it difficult to persuade the [decision-maker] that there is a real chance that the applicant will suffer persecution in the more distant future. But if the [decision-maker] is to apply the correct test ... it may be necessary to consider whether the applicant's fear of being persecuted in the more distant future (and not necessarily in the period shortly after his or her return) is well-founded', *NAGT of 2002 v Minister for Immigration*

an event happening upon return parallels assessments of whether expulsion would engage the host state's *non-refoulement* obligations under international human rights law, where standards such as 'imminence' find expression. But to read imminence into international refugee law is to deny the distinctiveness of the refugee predicament.

That the event paradigm risks losing sight of that distinctiveness is clearly illustrated by Hathaway and Pobjoy's characterisation of the predicament of Anne Frank developed in their critique of the judgment of the UK Supreme Court in *HJ and HT*.[68] In that case the Court unanimously accepted that a person would accurately be described as being persecuted for a Convention reason if, owing to a fear of harm, he chooses to conceal his non-heterosexual orientation, perhaps indefinitely.[69] For Hathaway and Pobjoy, that person would not accurately be described as being persecuted because if the hiding were successful, the risk of harm would not accrue. They make the argument with reference to the predicament of Anne Frank, who they say would not have been able to establish a well-founded fear of being persecuted for a Convention reason if she had been able to successfully hide out in the attic until the risk of being deported to the death camps had subsided.[70]

The reasoning warrants reproduction as it highlights the incongruity of the event paradigm. First, the authors quote from the argument advanced by the Secretary of State that:

> If [Anne Frank] escaped and claimed asylum, the question would be whether she faced *a real risk of persecution* on return. The real Anne Frank would have been a refugee because she obviously did and therefore her example may not be helpful. But if (improbably) it was found that on return to Holland she would successfully avoid detection by hiding in the attic, the answer to the first stage of inquiry would be that she was not at real risk of persecution by the Nazis. But the second stage would be to ask whether permanent enforced confinement in her attic would itself amount to persecution. If it would, she would be a refugee.[71]

Lord Collins is identified as the most forceful critic of this approach, as reflected in his view that:

> Simply to re-state the Secretary of State's argument shows that it is not possible to characterize it as anything other than absurd and unreal. It is plain that it remains the threat to Jews of the concentration camp and the gas chamber which constitutes the persecution.[72]

Hence, the question whether a person can have a well-founded fear of being persecuted if the risk does not accrue is answered in the negative by the Secretary of State, and in the

---

and Multicultural and Indigenous Affairs [2002] FCAFC 319, [22], cited in Hathaway and Foster, *Refugee Status* (n 14) 123.

[68] *HJ (Iran) v Secretary of State for the Home Department* [2010] UKSC 31, [2011] AC 596.

[69] See for example Lord Rodger at 648: 'If, on the other hand, the tribunal concludes that a material reason for the applicant living discreetly on his return would be a fear of the persecution which would follow if he were to live openly as a gay man, then, other things being equal, his application should be accepted. Such a person has a well-founded fear of persecution. To reject his application on the ground that he could avoid the persecution by living discreetly would be to defeat the very right which the Convention exists to protect'.

[70] James Hathaway and Jason Pobjoy, 'Queer Cases Make Bad Law' (2011) 44 NYU J Int'l L & Pol 315, 350.

[71] Ibid 349 (emphasis added).   [72] Ibid 350.

affirmative by Lord Collins. Disagreeing with Lord Collins, Hathaway and Pobjoy consider the argument advanced by the Secretary of State to be 'neither absurd nor unreal'.[73] They continue:

> To the contrary, if it were found that Anne Frank could have successfully avoided detection by hiding in the attic, then the Secretary of State is entirely correct to say that she was not at real risk of being sent to the concentration camps.[74]

'Persecution', then, happens at the concentration camp, not, as Lord Collins thought, the moment the threat of being sent to the concentration camps for being a Jew becomes a reality. It therefore follows that a person's fear of being persecuted cannot be 'well-founded' if, in fact, the person will not suffer such harm. It is correct that a person with a guaranteed hiding place in the midst of danger could not establish a real risk of being sent to a concentration camp, but this view unduly constrains the experience of being persecuted to the harm that a person would be exposed to in such an environment.

This book contends that such an argument reflects an artificially narrow temporal scope of being persecuted. As Section 5.3 develops, the predicament of being persecuted is best understood as a condition of existence, permeated by risk and potentially punctuated by acts of persecution or other serious denials of human rights that reflect and reinforce that predicament. On this approach, Anne Frank would accurately be described as being persecuted the moment conditions exposed her to a real chance of serious denials of human rights demonstrative of a failure of state protection. On this approach, finding a guaranteed hiding place would not change the fact that Anne Frank was persecuted, because the need for a hiding place in the first place is evidence that she was persecuted. She might avoid being subjected to a specific act of persecution, but the discriminatory social context that forces her to hide gives rise to the experience of being persecuted and thus having to hide in order to avoid exposure to feared acts of persecution.

5.3 Why a Predicament Approach to Determining the Existence of a Well-Founded Fear of Being Persecuted Is Appropriate

The definition at Article 1A(2) of the Refugee Convention does not refer to a real risk of persecution. Rather, in both the English and French-language versions, the condition feared is that of being persecuted.[75]

The RSAA and its successor the NZIPT have consistently adopted an approach to the being persecuted element of the refugee definition that finds significance in the Convention's use of the passive, as opposed to the active, voice.[76] As noted in *Refugee Appeal No. 74665/03*:

> While it is common in refugee discourse to refer to 'the persecution element' of the refugee definition, the Authority prefers to use the language of the Convention itself, namely 'being persecuted'. Not only is this mandated by principles of treaty interpretation, it also serves to emphasise the employment of the passive voice. The inclusion clause has as its focus the predicament of the refugee claimant. The language draws attention to the fact of exposure to harm rather than to the act of inflicting harm.[77]

---

[73] Ibid [74] Ibid [75] In the French '*craignant avec raison d'être persécutée*', as noted above.
[76] See Section 3.2 for references. [77] *Refugee Appeal No. 74665/03* RSAA (7 July 2004), [36].

However, notwithstanding the express and principled adoption of a predicament approach, the RSAA remain preoccupied with acts of persecution, citing Hathaway with approval:

> [R]efugee law ought to concern itself with *actions which deny human dignity* in any key way and that the sustained or systemic denial of core human rights is the appropriate standard. In other words, core norms of international human rights law are relied on to define forms of serious harm within the scope of 'being persecuted'.[78]

This apparent incongruity between a predicament approach that simultaneously emphasises the conduct of actors of persecution can perhaps be explained by recognising that doctrinal and judicial development of the predicament approach to the nexus clause has taken as its counterpoint the intention-based approach to the nexus clause, as described in Chapter 3, rather than addressing itself to questions of the temporal scope of the experience of being persecuted. Thus, the predicament approach as currently articulated, and the emphasis placed there on the use of the passive voice in the refugee definition, only goes some way towards articulating a predicament approach to the refugee definition as a whole. It does not displace the dominant event paradigm and, apart from cases of cumulative grounds persecution,[79] remains concerned with the moment when serious harm is inflicted, notwithstanding recent recognition in *DS (Iran)* of the enduring character of the predicament.[80]

In advancing an argument in favour of a 'predicament approach' to being persecuted, this section focuses on a different consequence of the use of the passive voice in the refugee definition: the implications for the temporal scope. Acknowledging that the passive voice directs attention to the predicament of the claimant, as distinct from the conduct of the actor of persecution, a predicament approach must go further by transcending the event paradigm.

The step performed here is to make a clear distinction between what the Qualification Directive clearly identifies as 'acts of persecution' and the predicament of being persecuted. At the same time, the predicament approach articulated here does not disregard the relevance of conduct of actors of persecution. Without relevant conduct, a person cannot be persecuted. But the focus is on the overall predicament of the claimant in a wider temporal frame, rather than the isolated events in which conduct amounts to an act of persecution. The notion of the interpersonal–structural violence continuum described in Chapter 2 encapsulates this wider frame.

Anne Frank, compelled to hide in the secret annex at Prinsengracht 263 in Amsterdam, was persecuted. She was persecuted because her civil status (Jew) engendered (a real chance of being exposed to) serious denials of her human rights (in her case arbitrary arrest, detention, denial of the right to education, denial of the highest attainable standard of health, and so forth). She did not have to wait until her arbitrary arrest and deportation to Westerbork camp and then to Auschwitz-Birkenau for that label to apply to her predicament.[81] The condition of existence itself may be characterised by serious denials of human

---

[78] Ibid [58] (emphasis added), citing Hathaway, *Refugee Status* (n 11), 108. It is interesting to note Hathaway's use of 'denial' as opposed to 'violation' here.
[79] See reference in Section 3.2.   [80] *DS (Iran)* (n 29) [126].
[81] Note that the Nazis denied Jews the right to education in Holland from 1941, indicating Anne's first experience of a serious denial of human rights occurred at that point. Source: British Library <www.bl.uk/learning/histcitizen/voices/info/annefrank/annefrank.html> accessed 15 April 2019.

rights, thus making any notion of risk assessment, even within the being persecuted element of the definition, superfluous. For example, living in hiding without access to education would itself amount to a serious denial of human rights and would therefore, because the condition of existence is clearly connected to her civil status, amount to being persecuted.

The relevant conduct in this connection would be found in those acts and omissions by state and non-state actors that, together, engender Anne's (real chance of being exposed to) serious denials of human rights. Such conduct would clearly include orders to round up Jews, decisions to deny access to employment or education, and so forth. Non-state actors, including for example people who would inform of Anne's whereabouts, contribute to the creation of the persecutory social environment that Anne inhabits and that engenders her experience of being persecuted as a condition of existence. This framing is different from cumulative grounds persecution because the focus there is on the question of when a certain threshold of severity of an accumulation of discriminatory measures will be attained, whereas here the focus is on the ways in which conduct engenders and/or exacerbates (a real chance of being exposed to) serious denials of human rights.

This interpretation of being persecuted as a condition of existence, as distinct from an isolated act or accumulation of measures, finds support in the principles of treaty interpretation set out at Article 31 of the VCLT. Recalling the methodology that requires interpretation to consider the ordinary meaning of the term in context and in light of the object and purpose of the treaty, the clear starting point is to consider the term itself within the definition at Article 1A(2). The *Oxford English Dictionary* defines 'persecution' as '[a]n instance or act of persecuting; an injurious act', suggesting that event-based approaches to being persecuted have touched upon the uncontroversial meaning of the term. However, subsequent definitions envisage both isolated acts as well as more persistent conduct. For example, the third definition includes 'persistent annoyance or injury; harassment; an instance of this'.[82] Hence, 'persecution' is capable of having both a narrow, event-based meaning, but also contains elements of the persistency identified in the first edition of *The Law of Refugee Status* and retained in the second edition. Examples of usage of the passive form of the verb also reflect this persistency in contexts of clear relevance to the refugee predicament:

- To the glorye of hys persecuted churche (1546)
- A long wail of anguish was rising from the persecuted all over France (in a publication relating to the Huguenots) (1867)
- Several settlements of the persecuted sect of the Doukobohrs are established there [*i.e.* in Canada] (1899)[83]

Notably, the examples reflecting a wider temporal scope are precisely those in which the passive voice is used. This observation supports the argument that there is a relevant difference between 'persecution' and 'being persecuted', and that the latter reflects a condition of existence with a wider temporal scope.

Recognition of the wider temporal scope of being persecuted, beyond limitation to specific acts of persecution, is required by the faithful application of Article 31 of the VCLT. The narrower temporal scope reflected in event-based approaches fails to account

---

[82] 'persecution, n.' *OED Online* (OUP June 2017) accessed 23 August 2017.
[83] 'persecuted, adj. and n.' *OED Online* (OUP December 2016) accessed 15 January 2017.

for the other definitions of 'persecution' and the examples of being persecuted provided above. Recalling Article 31(4), it would fall to the party advancing the narrower interpretation to establish that the parties intended this more restrictive, event-based definition to apply.

Hence, it would be appropriate when describing the temporal scope of being persecuted to use the formulation 'a condition of existence entailing (a real chance of being exposed to) serious denials of human rights demonstrative of a failure of state protection'.

As noted in Chapter 4, the RSAA asserted in *Refugee Appeal No. 74665/03* that '[t]he risk of the anticipated harm occurring is a related, but separate enquiry. It must occur downstream of the inquiry whether the anticipated harm can properly be described as "being persecuted"'.[84] This chapter has argued against that proposition. Risk, it is contended here, is an inherent feature of being persecuted. It therefore invites conceptual confusion to speak of a 'real risk' of 'persecution'.

It warrants emphasising that this chapter's rejection of the doctrinal and jurisprudential excision of risk from the predicament of being persecuted does not entail a rejection of the need to assess risk in the context of RSD. This point is significant. The reasoning of senior judges in leading refugee law jurisdictions that adopts notions such as 'real risk', 'reasonable likelihood', and 'real chance' provides important guidance for the assessment of whether a person has a well-founded fear of being persecuted on return to her country of origin or habitual residence. The distinction made here is that the assessment of risk is part and parcel of the assessment of the claimant's predicament in her country of nationality or former habitual residence. It is not a subsequent question to be answered after it has been determined that the harm to which the person fears being exposed amounts to being persecuted. Under the approach developed in this chapter, the risk assessment is the same as under existing principles of international refugee law, but the location of the risk assessment is within the notion of being persecuted, as distinct from being a separate assessment that takes place after determining whether what the claimant fears amounts to being persecuted.

In the vast majority of cases, the recalibrated approach to being persecuted that moves away from the event paradigm and towards consideration of the wider predicament of the claimant in a broader temporal perspective will have no impact whatsoever on the outcome of the claim. In most cases, individuals seeking recognition of refugee status do express a fear of being exposed to a particular kind of harm, typically at the hands of an identifiable actor of persecution. These scenarios, whether relating to a political dissident fearing arrest and torture, a lesbian woman fearing 'corrective' rape, or a religious convert fearing execution, are all readily accommodated under an interpretation of being persecuted as reflecting a condition of existence permeated by the risk of serious harm. Different outcomes are only likely to result in more complex cases reflecting the kind of structural violence described in Chapter 2, where actors of persecution are diffuse, the risk of serious denials of human rights is endemic, and state protection is routinely inadequate, including in the context of disasters and climate change.

However, as will become increasingly clear over the course of the coming two chapters, this broader temporal scope does very little indeed to extend eligibility for recognition of

---

[84] *Refugee Appeal No. 74665/03* (n 77) [125].

refugee status to the millions of people exposed to displacement and other forms of harm in the context of disasters and climate change. A person who faces return to a condition of existence entailing (a real chance of being exposed to) serious denials of human rights demonstrative of a failure of state protection is not, without more, equivalent to a person who faces being persecuted. The essential element of discrimination must also be established. This element is examined in the following chapter.

6

The Personal Scope of Being Persecuted: The Function of the Non-discrimination Norm within the Refugee Definition

6.1 The Non-discrimination Norm

The joined norms of equality and non-discrimination pervade international human rights treaties as well as domestic constitutional documents the world over. Although no systematic analysis appears to have been conducted to address the question whether the general prohibition on discrimination has attained the status of customary international law, there are some compelling judicial[1] and academic commentaries[2] endorsing such a proposition, over and above the specific prohibition on racial discrimination recognised in the *Barcelona Traction* case.[3]

With the exception of the Convention on Enforced Disappearances,[4] international human rights treaties afford primacy to the non-discrimination norm before elaborating more substantive rights. In addition to CEDAW[5] and CERD,[6] which address discrimination against women and discrimination on grounds of race, colour, descent, or national or ethnic origin respectively, general non-discrimination provisions are found in the ICCPR,[7] ICESCR,[8] CRC,[9] the Convention on Migrant Workers (CMW)[10] and the Convention on the Rights of Persons with Disabilities (CRPD).[11] Non-discrimination is also identified as a foundational principle in the UN Charter[12] and the Universal Declaration of Human Rights (UDHR).[13]

---

[1] *Legal Consequences for States of the Continued Presence of South Africa in Namibia (South West Africa) notwithstanding Security Council Resolution 276 (1970)* (Advisory Opinion) [1971] ICJ Rep 16, 76 (Separate Opinion of Judge Ammoun); Inter-American Court of Human Rights, Advisory Opinion OC-18/03 of 17 September 2003, Juridical Condition and Rights of Undocumented Migrants, [101].

[2] Richard B Lillich, 'Civil Rights' in T Meron (ed), *Human Rights in International Law: Legal and Policy Issues* (OUP 1984).

[3] *Barcelona Traction, Light and Power Company, Limited* (Judgment) [1970] ICJ Rep 3, 32.

[4] International Convention for the Protection of All Persons from Enforced Disappearance (adopted 20 December 2006, entered into force 23 December 2010) 2716 UNTS 3.

[5] See for example Article 1, Convention on the Elimination of All Forms of Discrimination Against Women (adopted 18 December 1979, entered into force 3 September 1981) 1249 UNTS 13.

[6] See for example Article 1, International Convention on the Elimination of All Forms of Racial Discrimination (adopted 21 December 1965, entered into force 4 January 1969) 660 UNTS 195.

[7] Articles 2(1) and 26 ICCPR.   [8] Article 2(2) ICESCR.   [9] Article 2(1) CRC.

[10] Article 7, International Convention on the Protection of the Rights of All Migrant Workers and Members of their Families (adopted 18 December 1990, entered into force 1 July 2003) 2220 UNTS 3

[11] Article 5, Convention on the Rights of Persons with Disabilities (adopted 13 December 2006, entered into force 3 May 2008) 2515 UNTS 3.

[12] Article 1 United Nations, Charter of the United Nations (24 October 1945) 1 UNTS XVI.

[13] Article 2 UDHR.

Article 1 CERD prohibits 'racial discrimination', defined as:

> Any distinction, exclusion, restriction or preference based on race, colour, descent, or national or ethnic origin which has the purpose or effect of nullifying or impairing the recognition, enjoyment or exercise, on an equal footing, of human rights and fundamental freedoms in the political, economic, social, cultural or any other field of public life.

There are 179 states that are party to CERD.[14]

A similar definition focusing on discrimination against women is provided at Article 1 CEDAW, which has 189 States Parties.[15] Referencing both definitions, the HRC defines discrimination for the purposes of the ICCPR as:

> Any distinction, exclusion, restriction or preference which is based on any ground such as race, colour, sex, language, religion, political or other opinion, national or social origin, property, birth or other status, and which has the purpose or effect of nullifying or impairing the recognition, enjoyment or exercise by all persons, on an equal footing, of all rights and freedoms.[16]

There are 172 states that are party to the ICCPR.[17]

Article 2 of the Convention on the Rights of the Child prohibits discrimination:

> States Parties shall respect and ensure the rights set forth in the present Convention to each child within their jurisdiction without discrimination of any kind, irrespective of the child's or his or her parent's or legal guardian's race, colour, sex, language, religion, political or other opinion, national, ethnic or social origin, property, disability, birth or other status.

There are 196 states that are party to the CRC.[18]

Without purporting to assert the existence of a customary norm of general international law prohibiting discrimination in all of its forms, in light of the foregoing there is at least cause to consider the possibility that such a norm does exist. Whether the general non-discrimination norm has attained customary status or not, a pragmatic approach to interpretation, similar to Hathaway's invocation of the 'super-majority' principle,[19] would find merit in having regard to the non-discrimination norm as a relevant rule of international law applicable in the relations between the parties, following Article 31(3)(c) VCLT, when interpreting the refugee definition. Indeed, this chapter argues that the non-discrimination norm acts as a central organising principle around which the refugee definition is to be interpreted.

So how is the non-discrimination obligation under treaty law to be understood? The person who faces discrimination need not actually possess the characteristic that is the basis for the discrimination. It is enough that this characteristic is imputed to her.[20] Of equal

---

[14] See United Nations Treaty Collection, <https://treaties.un.org/Pages/ViewDetails.aspx?src=IND&mtdsg_no=IV-2&chapter=4&clang=_en> accessed 23 February 2019.

[15] See United Nations Treaty Collection, <https://treaties.un.org/Pages/ViewDetails.aspx?src=IND&mtdsg_no=IV-8&chapter=4&lang=en> accessed 23 February 2019.

[16] HRC, 'General Comment No. 18: Non-discrimination' (10 November 1989) [7].

[17] See United Nations Treaty Collection, <https://treaties.un.org/Pages/ViewDetails.aspx?chapter=4&clang=_en&mtdsg_no=IV-4&src=IND> accessed 23 February 2019.

[18] See United Nations Treaty Collection, <https://treaties.un.org/pages/ViewDetails.aspx?src=IND&mtdsg_no=IV-11&chapter=4&lang=en> accessed 23 February 2019.

[19] See reference to this super-majority principle in Section 3.2.

[20] CESCR, 'General Comment No. 20: Non-Discrimination in Economic, Social and Cultural Rights (Art. 2, Para. 2, of the International Covenant on Economic, Social and Cultural Rights)' (2 July 2009) E/C.12/GC/20, [16].

importance is the recognition that discrimination may take place with or without a discriminatory intent,[21] and may result from the conduct of both state and non-state actors.[22] As recognised by CERD, ICESCR, CEDAW, and CRC, discrimination may take place on multiple grounds, with gender-based discrimination often recognised as interacting with other grounds such as race.[23]

General Comment No. 20 by the Committee on Economic, Social and Cultural Rights provides detailed treatment of the forms that discrimination can take, and the taxonomy is very similar across other international human rights instruments.[24] Discrimination may be direct, in the sense that legal provisions or the conduct of relevant actors specifically single out persons for differential treatment on prohibited grounds. Discrimination may also be indirect, in the sense that particular groups are disproportionately affected by certain, apparently neutral, measures or omissions.[25] States have obligations to take steps to eradicate both de jure discrimination as well as de facto discrimination. States have both positive and negative obligations in relation to the non-discrimination norm. Hence, states must take steps not only to respect rights to non-discrimination by, for example, refraining from passing discriminatory legislation and amending or repealing existing discriminatory legislation, but also to protect people within their jurisdiction from the discriminatory conduct of private actors, and to take steps towards the fulfilment of the right to non-discrimination, for example through public education campaigns.[26]

De jure discrimination, whether direct or indirect, must be eradicated with immediate effect. De facto discrimination 'should be brought to an end as speedily as possible'.[27] Structural or systemic inequality is a particularly difficult form of discrimination to eradicate as it arises as a result of the influence of 'dominant social values' and 'a State's prevalent race, religion, and language'. According to Joseph and Castan:

> The reinforcement of dominant norms generates systemic discrimination against, and consequent systemic inequality for, people outside the dominant norm. The causes and effects of systemic inequality may be very subtle and even invisible within parts of society, as such inequality is historically perceived as 'normal' by both the dominant and marginalized people.[28]

The recognition of some degree of 'progressive realisation' inherent in the obligation to eliminate de facto discrimination stems from the recognition that certain patterns of discrimination may be deeply entrenched. Such 'systemic' discrimination receives particular attention in the CESCR's General Comment No. 20:

---

[21] Wouter Vadenhole, *Non-discrimination and Equality in the View of the UN Human Rights Treaty Bodies* (Intersentia 2005) 35. Cases reflecting this position include *Zwaan de Vries v the Netherlands* CCPR/C/29/D/182/1984; *Simunek et al. v Czech Republic* (CCPR/C/54/D/516/1992).

[22] Ibid 85. Vadenhole identifies CERD, 'Concluding Observations of the Committee on the Elimination of Racial Discrimination: Brazil' (28 April 2004) CERD/C/64/CO/2 concerning de facto racial segregation in slum areas (*favelas*) as an example of where CERD has recognised discrimination by non-state actors.

[23] Ibid   [24] Vadenhole, *Non-discrimination and Equality* (n 21) 34–36.

[25] See for example *L.R. and Others (represented by European Roma Rights Centre) v Slovakia*, CERD, 5 August 2003, Communication No. 31/2003 and *DH and Others v Czech Republic* [2008] 47 EHRR 3.

[26] See extensive discussion of states' obligations to respect, protect, and fulfil the right to non-discrimination in Vadenhole, *Non-discrimination and Equality* (n 21) ch. 4.

[27] The Limburg Principles on the Implementation of the International Covenant on Economic, Social and Cultural Rights (adopted 8 January 1987) UN Doc. E/CN.4/1987/17/, [38].

[28] Sarah Joseph and Melissa Castan, *The International Covenant on Civil and Political Rights: Cases, Materials and Commentary* (3rd edn, OUP 2013) 817.

The Committee has regularly found that discrimination against some groups is pervasive and persistent and deeply entrenched in social behavior and organization, often involving unchallenged or indirect discrimination. Such systemic discrimination can be understood as legal rules, policies, practices or predominant cultural attitudes in either the public or private sector which create relative disadvantage for some groups and privileges for other groups.[29]

However, notwithstanding the enormous challenges associated with eradicating systemic discrimination, the CESCR does not recognise unlimited flexibility as to the realisation of this core objective:

A failure to remove differential treatment on the basis of a lack of available resources is not an objective and reasonable justification unless every effort has been made to use all resources that are at the State party's disposition in an effort to address and eliminate the discrimination, as a matter of priority.[30]

However, as Joseph and Castan note, 'it is doubtful that systemic inequality could be successfully challenged before the HRC in an Optional Protocol complaint',[31] and therefore more attention has been paid to systemic discrimination within the periodic review processes under individual treaty monitoring mechanisms.[32]

Particularly in relation to systemic discrimination, it warrants emphasising that a person may experience discrimination even when the state is making 'every effort'[33] to use 'all resources that are at the State party's disposition ... as a matter of priority',[34] most clearly in situations where the discrimination emanates from 'predominant cultural attitudes in either the public or private sector which create relative disadvantage',[35] not least when multiple discrimination is manifest. The similarity between the legal concept of systemic discrimination and the earlier described concept of structural violence should be highlighted at this stage and will be developed further in Chapter 7.

6.2 The Non-discrimination Norm in International Refugee Law

In order to comply with Article 31 of the VCLT, the interpretation of the refugee definition at Article 1A(2) of the Refugee Convention must take account of the non-discrimination norm. First, the non-discrimination norm is part of the context. Just as reference to human rights in the preamble is identified as requiring consideration of international human rights law (IHRL), so too must the reference to the principle of non-discrimination in the same

---

[29] CESCR, 'General Comment No. 20: Non-discrimination in Economic, Social and Cultural Rights (Art. 2, Para. 2, of the International Covenant on Economic, Social and Cultural Rights)' (2 July 2009) E/C.12/GC/20 [12].
[30] Ibid [13].
[31] Joseph and Castan, *Cases, Materials and Commentary* (n 28) 818. Vadenhole agrees expressly with this point. See Vadenhole, *Non-discrimination and Equality* (n 21) 61.
[32] See for example CCPR/CO/83/KEN (women in Kenya); CCPR/C/79/Add.81 (women in low castes in India); E/12/1/Add.92 (women and girls in Yemen); E/12/1/Add.61 (ethnic minorities in Togo); E/12/1/Add.93 (indigenous people in Guatemala); CRC/C//15/Add.246 (children in Angola), all cited (along with many others) in Vadenhole, *Non-discrimination and Equality* (n 21).
[33] CESCR, 'General Comment No. 20: Non-discrimination in Economic, Social and Cultural Rights' (n 29) [13].
[34] Ibid   [35] Ibid [12].

perambular paragraph influence how the refugee definition is interpreted. Second, the compound notion of 'being persecuted for reasons of race, religion, nationality, membership of a particular social group or political opinion' closely resembles non-discrimination provisions in the IHRL instruments surveyed above. Third, recalling the argument developed in relation to Article 31(3)(c) and the fact that a 'super-majority' of states have agreed to be bound by some iteration of the non-discrimination norm, the non-discrimination norm is a relevant rule of international law applicable in the relations between the parties that shall be taken into account. Finally, as this part of the chapter develops, discrimination, far more than substantive human rights norms, is inherent in the ordinary meaning of being persecuted, cementing non-discrimination as the central organising principle around which the refugee definition falls to be interpreted.

In advancing this argument, inspiration is drawn from Cantor's identification of a basis for developing 'a new theoretical model of refugee law' that foregrounds discrimination:

> By focusing first on the discrimination, rather than the harm to which it leads, this approach seeks to avoid putting the cart before the horse ... Rather than being a human rights violation that certain groups of putative refugees cannot remedy in the country of origin, persecution is instead viewed as an exacerbated form of discrimination that follows from the making of unjust distinctions recognised by the Convention grounds ... [T]his approach offers an intriguing basis for researching into the development of a new theoretical model of refugee law in the future to supersede the innovative and influential model proposed by Hathaway in 1991.[36]

This chapter embraces Cantor's call for a 'new theoretical model' built around the non-discrimination norm, and by way of conclusion attempts a more thoroughgoing recalibration of the interpretation of the refugee definition as a whole.

In order to adequately reflect the non-discrimination norm, any interpretation of the refugee definition at Article 1A(2) must address not only direct discrimination, but also other forms of discrimination described above. Thus, the interpretation of being persecuted for a Convention reason must also accommodate indirect and systemic discrimination, both de facto and de jure.

It is not controversial that the notion of discrimination plays some kind of role in the refugee definition. However, the event paradigm reflected in the definition of being persecuted as the sustained or systemic denial of human rights demonstrative of a failure of state protection locates the discriminatory element of the refugee definition within the nexus clause, whilst also permitting discriminatory conduct as examples of acts of persecution. In contrast, the predicament approach developed in Chapter 5 would recognise the discriminatory element as integral to the notion of 'being persecuted'. In what follows, the role of discrimination within the event paradigm is presented first, followed by a discussion of approaches that have identified discrimination as an integral feature of being persecuted.

---

[36] David Cantor, 'Defining Refugees: Persecution, Surrogacy and the Human Rights Paradigm' in Bruce Burson and David Cantor (eds), *Human Rights and the Refugee Definition: Comparative Legal Practice and Theory* (Brill Nijhoff 2016) 392–94.

6.2.1 The Non-discrimination Norm Reflected in the Nexus Clause

In considering the potential of non-discrimination principles to explain 'not only the scope of the refugee definition but also, potentially, the nature of the Convention protection regime as a whole', Cantor locates discrimination firmly within the Convention grounds:

> [A]gainst the persecution-centric approach of many human rights-based models, another alternative identifies the principle of non-discrimination as the basis for refugee law. It suggests that the Hathaway model's overemphasis on the harm implicit in the persecution element obscures the principle of non-discrimination expressed by the Convention grounds as a more productive starting point for legal interpretation of the refugee definition.[37]

In recognising the principle of non-discrimination 'expressed in the Convention grounds', as distinct from a 'persecution-centric approach', Cantor appears to suggest that discrimination is distinct from 'persecution'. Hence, notwithstanding his call for a new theoretical approach to the refugee definition that is based on the non-discrimination obligation, he continues to conceive of being persecuted from within the event paradigm.

Chetail also locates the non-discrimination element clearly within the Convention grounds:

> The grounds of persecution provide one of the most obvious instances of the human rights filiation: the grounds of religion and political opinion are clearly based on freedom of thought and that of opinion and expression, while the other ones – race, nationality, and membership of a particular social group – are anchored within the principle of non-discrimination.[38]

Why Chetail does not also associate being persecuted for reasons of political opinion and religious belief with the non-discrimination obligation is unclear, given these are clearly incorporated into the prohibited grounds under many instruments.[39]

The fact that the Convention grounds are reflected in non-discrimination provisions under IHRL draws attention to the nexus between being persecuted and the Convention grounds as the site at which the discrimination that is integral to the refugee definition must be identified in an individual case. This approach is clearly reflected in the methodology adopted by the NZIPT. As noted in Chapter 3, since *Refugee Appeal No. 70074*,[40] the Refugee Status Appeals Authority and its successor the NZIPT have applied the following methodology for determining claims for refugee status:

- Is there a well-founded fear of being persecuted?
- Is there a connection to one or more of the Convention reasons?

As the test articulated by the NZIPT highlights, the question whether a person's well-founded fear of being persecuted is sufficiently proximate to a Convention reason falls to be determined only after it has been established that the person has a well-founded fear of

---

[37] Ibid 393.
[38] Vincent Chetail, 'Are Refugee Rights Human Rights? An Unorthodox Questioning of the Relations between Refugee Law and Human Rights Law' in Ruth Rubio-Marin (ed), *Human Rights and Immigration* (OUP 2014) 27 (references omitted).
[39] Consider Article 2 UDHR, Article 2(1) ICCPR, and Article 2(2) ICESCR.
[40] *Refugee Appeal No. 70074/96* RSAA (17 September 1996).

being persecuted,[41] where being persecuted is understood as the sustained or systemic violation of human rights demonstrative of a failure of state protection. Having established the claim to this extent, the second question that falls to be determined is whether the fear of being persecuted is 'for reasons of' a person's race, religion, nationality, membership of a particular social group, or political opinion.

The problem with this approach is that it takes the human-rights-based interpretation of the refugee definition so far that the distinctive quality of being persecuted is diluted, with the effect that being persecuted becomes indistinguishable from being the victim of a human rights violation. The result is that the category of people who can accurately be described as being persecuted becomes unduly broad, encompassing all persons who are exposed to a sustained or systemic violation of human rights demonstrative of a failure of state protection, including all victims of 'natural' disasters that result from failures by the state to fulfil its positive obligations to protect the right to life, health, property, and so forth from foreseeable hazards.

6.2.2 The Non-discrimination Norm Reflected in the Notion of Being Persecuted

The event paradigm does not, however, isolate the discrimination within the nexus clause. Rather, it also recognises certain *discriminatory measures* as potentially amounting to acts of persecution. For example, the Qualification Directive identifies 'prosecution or punishment which is disproportionate or discriminatory' and 'denial of judicial redress resulting in a disproportionate or discriminatory punishment'[42] as acts of persecution, suggesting that it is the discriminatory element that transforms the conduct into an act of persecution. Pursuant to Article 9(3), there must nevertheless be established 'a connection' between the acts of persecution and the Convention reasons.[43] Hence, discriminatory prosecution must establish a connection with the grounds of discrimination reflected in the Convention reasons, suggesting the adjective is superfluous in relation to the act.

In other cases, the focus is more specifically on discrimination in relation to the enjoyment of economic and social rights. The classic statement of this position is found in the UNHCR Handbook:

> 54. Differences in the treatment of various groups do indeed exist to a greater or lesser extent in many societies. Persons who receive less favourable treatment as a result of such differences are not necessarily victims of persecution. It is only in certain circumstances that discrimination will amount to persecution. This would be so if measures of discrimination lead to consequences of a substantially prejudicial nature for the person

---

[41] See discussion in Section 3.2 regarding conceptual challenges to dominant interpretations of these (often interconnected) elements of the refugee definition.

[42] Council of the European Union, Directive 2011/95/EU of the European Parliament and of the Council of 13 December 2011 on Standards for the Qualification of Third-Country Nationals or Stateless Persons as Beneficiaries of International Protection, for a Uniform Status for Refugees or for Persons Eligible for Subsidiary Protection, and for the Content of the Protection Granted (Recast), 20 December 2011, OJ L. 337/9-337/26; 20.12.2011 (Qualification Directive) Article 9(2)(c)–(d).

[43] 'In accordance with point (d) of Article 2, there must be a connection between the reasons mentioned in Article 10 and the acts of persecution as qualified in paragraph 1 of this Article or the absence of protection against such acts.'

concerned, e.g. serious restrictions on his right to earn his livelihood, his right to practise his religion, or his access to normally available educational facilities.[44]

The recognition that discrimination can engender serious denials of the right to work, the right to freedom of thought, conscience, or religion, and the right to education, and that the exposure to the denial of such rights can reflect the predicament of being persecuted is an important feature of the predicament approach developed below. The Handbook diverges from that approach, however, by suggesting that discrimination is only a potential feature of 'persecution'. This position is somewhat surprising, as the Handbook appears to understand 'being persecuted for a Convention reason' as a compound notion:

> From Article 33 of the 1951 Convention it may be inferred that a threat to life or freedom on account of race, religion, nationality, political opinion or membership of a particular social group is always persecution. Other serious violations of human rights – for the same reasons – would also constitute persecution.[45]

Grahl-Madsen, despite being influenced by the event paradigm, reveals a perspective that sees discrimination as inherent to being persecuted. He discusses the relevance of the UDHR to the interpretation of 'persecution':

> It also seems natural to speak of persecution if, *for political or related reasons*, a person is subjected to slavery [Article 4]; to torture or to cruel, inhuman or degrading treatment or punishment [Article 5]; or to arbitrary arrest, detention or exile [Article 9].[46]

He later notes that '[r]elegation to a sub-standard dwelling place may be considered as persecution, but here again emphasis must be put on the political motivation of the measure'.[47]

The idea that discrimination is inherent to the experience of being persecuted is explained clearly by McHugh J:

> Persecution for a Convention reason may take an infinite variety of forms from death or torture to the deprivation of opportunities to compete on equal terms with other members of the relevant society. *Whether or not conduct constitutes persecution in the Convention sense does not depend on the nature of the conduct. It depends on whether it discriminates against a person because of race, religion, nationality, political opinion or membership of a social group.*[48]

Although Chapter 5 has challenged approaches that equate 'persecution' with the acts that reflect the condition of existence of being persecuted, the recognition that 'persecution' cannot be understood without reference to discrimination is important.

---

[44] UNHCR, *Handbook and Guidelines on Procedures and Criteria for Determining Refugee Status under the 1951 Convention and the 1967 Protocol Relating to the Status of Refugees* (Reissued December 2011) [54] <http://www.unhcr.org/publications/legal/3d58e13b4/handbook-procedures-criteria-determining-refugee-status-under-1951-convention.html> accessed 4 April 2019.
[45] Ibid [51].
[46] Atle Grahl-Madsen, *The Status of Refugees in International Law* (AW Sijthoff's Uitgeversmaatschappij NV 1966) 195.
[47] Ibid 215.
[48] *Applicant A v Minister for Immigration and Ethnic Affairs* [1997] HCA 4, [1997] 190 CLR 225, 258 (McHugh J) (emphasis added).

In another judgment by the High Court of Australia, McHugh J articulates a general definition of 'persecution' for the purposes of the Refugee Convention, in which discrimination is inherent:

> Ordinarily, however, given the rationale of the Convention, persecution for that purpose is:
>
> - unjustifiable and discriminatory conduct directed at an individual or group for a Convention reason
> - which constitutes an interference with the basic human rights or dignity of that person or the persons in the group
> - which the country of nationality authorises or does not stop, and
> - which is so oppressive or likely to be repeated or maintained that the person threatened cannot be expected to tolerate it, so that flight from, or refusal to return to, that country is the understandable choice of the individual concerned.[49]

The same perspective is articulated by Goudron J in the same case, who describes discrimination as 'an essential feature of persecution for the purposes of the Convention'.[50]

In *Ram*, the Federal Court of Australia explains:

> Persecution involves the infliction of harm, but it implies something more: an element of an attitude on the part of those who persecute which leads to the infliction of harm, or an element of motivation (however twisted) for the infliction of harm. People are persecuted for something perceived about them or attributed to them by their persecutors.[51]

A similar view is expressed by Lord Millett in *Islam*:

> The 1951 Convention was concerned to afford refuge to the victims of certain kinds of *discriminatory persecution*, but it was not directed to prohibit discrimination as such nor to grant refuge to the victims of discrimination.[52]

Different approaches to the relevance of the intention of the actor of persecution are discernible between these formulations that situate discrimination as an inherent feature of 'persecution'. However, they are united in their expression of the view that discrimination is indeed central to the definition.[53]

When being persecuted for a Convention reason is understood as a compound notion, which is appropriate in light of the injunctions from Sedley LJ and McHugh J at the beginning of Chapter 4, discrimination must take a central role in its interpretation. However, when it is necessary to consider the elements in isolation, the discrimination inherent in being persecuted for a Convention reason lies clearly in the verb, with the Convention reasons serving only to delimit the prohibited grounds.

---

[49] *Minister for Immigration v Haji Ibrahim* [2000] HCA 55, [2000] 204 CLR 1 [65]. [50] Ibid [29].

[51] *Ram v Minister for Immigration and Ethnic Affairs*, [1995] 130 ALR 213 (Aus. FFC, Jun. 27, 1995) at 571, cited in Michelle Foster, *International Refugee Law and Socio-Economic Rights: Refuge from Deprivation* (CUP 2007) 272.

[52] *Islam v Secretary of State for the Home Department Immigration Appeal Tribunal and Another, ex parte Shah, R v* [1999] UKHL 20, [1999] AC 629, 660 (emphasis added).

[53] Further support for the proposition that discrimination is inherent to being persecuted may be found in JF Durieux, 'Of War, Flows, Laws and Flaws: A Reply to Hugo Storey' (2012) 31 REFUG SURV Q 161, 165.

In what follows, arguments supporting the recognition of discrimination as an inherent feature of the experience of being persecuted are developed before challenges to the approach are considered.

6.3 Discrimination as an Inherent Feature of Being Persecuted

The occurrence of an act of persecution is evidence that a person has a well-founded fear of being persecuted, because being exposed to denials of human rights is a feature of the predicament of a person who is persecuted. However, to focus on the forms of harm, and the acts that cause such harm, is to miss what is distinctive about the legal character of being persecuted. It misses the fact that, as noted multiple times in this book, many people may be victims of a sustained or systemic denial of human rights demonstrative of a failure of state protection, but will not accurately be described as being persecuted within the meaning the term has within the Refugee Convention. The fact that the class of person encompassed by this definition is, by any measure, far wider than the class of person falling within the scope of the refugee definition as a whole, represents a limitation of the prevailing human-rights-based approach to being persecuted.

The interpretation that incorporates discrimination as an inherent feature of being persecuted is faithful to Article 31(1) of the VCLT. Applying the 'crucible approach' that recognises the ordinary meaning in context and in light of the object and purpose of the treaty as a single interpretative act,[54] a sensible place to begin is, nevertheless, with what Linderfalk terms 'conventional language'.[55] Indeed, Einarsen's observation that the drafters 'reckoned persecution to be a well-known concept'[56] warrants bearing in mind.

The *Oxford English Dictionary* defines 'to persecute' as:

> To seek out and subject (a person, group, organization, etc.) to hostility or ill-treatment, esp. on grounds of religious faith, political belief, race, etc.; to torment; to oppress.[57]

Thus, on this definition, being persecuted would entail the experience of being sought out and subjected to hostility or ill-treatment on account of one's civil or political status.[58] Discrimination is integral to the definition.

---

[54] Malagosia Fitzmaurice, 'Interpretation of Human Rights Treaties' in Dinah Shelton (ed), *The Oxford Handbook of International Human Rights Law* (OUP 2013) 746–47.

[55] Ulf Linderfalk, *On the Interpretation of Treaties: The Modern International Law as Expressed in the 1969 Vienna Convention on the Law of Treaties* (Springer 2010) 95: 'If it can be shown that in a treaty provision, there is an expression whose form corresponds to an expression of conventional language, then the provision shall be understood in accordance with the rules of that language'. Agreeing with leading scholars of international refugee law that the drafters probably intentionally left the term 'being persecuted' undefined so as not to unduly delimit the meaning of the term (see discussion in Section 3.2), 'conventional language' is taken in the case of the Refugee Convention to include contemporary usage. However, as will be demonstrated below, the conventional usage of the term has not changed significantly in hundreds of years.

[56] Terje Einarsen, 'Drafting History of the 1951 Convention and the 1967 Protocol' in A Zimmermann (ed), *The 1951 Convention Relating to the Status of Refugees and Its 1967 Protocol: A Commentary* (OUP 2011) 56–57.

[57] 'persecute, v.' *OED Online* (OUP December 2016) accessed 15 January 2017.

[58] As explained in Section 3.2, although there is not a unitary understanding of the term, there are cogent and widely recognised reasons for rejecting the notion that a claimant must demonstrate that they have

As already noted in Chapter 5, as a way of emphasising the extended temporal scope of being persecuted, the *Dictionary* provides a number of citations from texts dating back to the sixteenth century where the passive form 'being persecuted' has been employed with reference to race, religion, or other characteristics now protected by the non-discrimination norm. The citations are repeated here for ease of reference:

- To the glorye of hys persecuted churche (1546)
- A long wail of anguish was rising from the persecuted all over France (in a publication relating to the Huguenots) (1867)
- Several settlements of the persecuted sect of the Doukobohrs are established there [i.e. in Canada] (1899)[59]

However, the *Dictionary* also provides definitions under which no reference is made to discrimination:

- To afflict, trouble, vex, worry; to harass; to pester, importune, or annoy persistently
- To chase, hunt, or pursue with intent to capture, injure, or kill
- To follow up, pursue, prosecute (a subject); to carry out, go through with
- To prosecute (a person or †suit) at law. Now regional and humorous.[60]

There are also multiple examples in which 'persecuted' is used in a way that does not relate to any grounds of discrimination:

- The deep-mouth'd Hound ... following still. The persecuted Creature, to, and fro. (1697)
- The poor persecuted creature to which I allude is the Hedge-hog or Urchin. (1779)[61]

Similar definitions and examples of both 'discriminatory persecution' and persecution that does not include a discriminatory element can be found in the French language.[62]

Recourse to dictionary definitions is rightly repudiated as being in itself insufficient to reveal the meaning of the term in context and in light of the object and purpose of the treaty.[63] However, considered in relation to the immediate context of Article 1A(2), particularly the nexus clause and Convention grounds, as well as the non-discrimination reference in the preamble and the overarching norm of non-discrimination as a relevant rule of international law applicable in the relations between the parties, it is clear that the definition that comes closest to incorporating the element of discrimination is the most relevant to an understanding of the ordinary meaning.

Further judicial support for the proposition that discrimination is an inherent feature of being persecuted can be derived from the Opinion of Lord Hoffman in *Islam*:

---

been singled out for ill treatment in order to establish that their experience of being exposed to serious harm on account of their civil or political status amounted to being persecuted.

[59] 'persecuted, adj. and n.' *OED Online* (OUP December 2016) accessed 15 January 2017.
[60] 'persecute, v.' *OED Online* (OUP September 2016) accessed 7 October 2016.
[61] 'persecuted, adj. and n.' *OED Online* (OUP December 2016) accessed 15 January 2017.
[62] 'persécuter, verbe transitif' Larousse <http://www.larousse.fr/dictionnaires/francais/persécuter/59777> accessed 15 April 2019.
[63] See for example *Minister for Immigration and Multicultural Affairs v Khawar* [2002] HCA 14, [2002] 210 CLR 1 [108] (Kirby J) and *Refugee Appeal No. 71427/99* RSAA (16 August 2000) [46] (Roger Haines QC). See also James Hathaway and Michelle Foster, *The Law of Refugee Status* (2nd edn, CUP 2014) 190.

In my opinion, the concept of discrimination in matters affecting fundamental rights and freedoms is central to an understanding of the Convention. It is concerned not with all cases of persecution, even if they involve denials of human rights, but with persecution which is based on discrimination. And in the context of a human rights instrument, discrimination means making distinctions which principles of fundamental human rights regard as inconsistent with the right of every human being to equal treatment and respect. The obvious examples, based on the experience of the persecutions in Europe which would have been in the minds of the delegates in 1951, were race, religion, nationality and political opinion.[64]

Particularly when read in light of the different definitions of the verb 'to be persecuted' identified above, it is clear that Lord Hoffman firmly situates discrimination within the 'being persecuted' element of the refugee definition by first referring to the relevant, discriminatory, form of persecution amongst different forms, and second by locating that notion of discriminatory persecution within the wider framework of IHRL.

The opposite reading of Lord Hoffman is articulated by the NZIPT in *DS (Iran)*, in which an argument accepted by the High Court of New Zealand (NZHC) equating being persecuted with being discriminated against relied expressly on Lord Hoffman's words cited above. The NZHC read Lord Hoffman thus:

> The above statements of principle display acceptance of a linkage between persecution and discrimination. The second respondent has argued that 'persecution is a strong word' and that there is a 'clear distinction between a breach of human rights (discrimination) and a sustained or systemic denial of core human rights (persecution)'. The second respondent seemingly attempts to distinguish one from the other on the basis that discrimination is a less serious intrusion on human rights than is persecution. However, this argument overlooks the meaning ascribed to 'discrimination' by Lord Hoffman in *ex parte Shah*, who uses the term to refer to circumstances where targeted individuals are stripped of the enjoyment of fundamental human rights. It is the targeted and therefore discriminatory nature of such ill-treatment that leads to it being persecution.[65]

The NZIPT rejected this reading, arguing that Lord Hoffman 'was not in fact referring to the "being persecuted" limb of the refugee definition':

> Rather than describing the 'being persecuted' element, Lord Hoffman was clearly referencing the 'nexus to a Convention ground' element of the refugee definition and the underlying concern of the drafters to provide a regime of international protection for persons being discriminated against in the enjoyment of their human rights by reason of their having one of the Convention-protected characteristics.

The Tribunal continues:

> Discrimination in the enjoyment of human rights by reason of possession of a Convention-protected characteristic constitutes a breach of the underlying anti-discrimination norm upon which the nexus element of the refugee definition rests, but does not, of itself, constitute being persecuted.[66]

---

[64] *Islam* (n 52) 651 (Lord Hoffman).
[65] *CV v Immigration and Protection Tribunal and CW v Immigration and Protection Tribunal* [2015] NZHC 510, [103], cited in *DS (Iran)* NZIPT 800788 [136].
[66] *DS (Iran)* (n 65) [139]. The context in which the NZIPT makes this observation is relevant. *DS (Iran)* was concerned with an argument that the notion of being persecuted did not require evidence that the denial

The better view is that Lord Hoffman was referring to the being persecuted element when discussing the centrality of discrimination. Later in his judgment, he explains further:

> Assume that during a time of civil unrest, women are particularly vulnerable to attack by marauding men, because the attacks are sexually motivated or because they are thought weaker and less able to defend themselves. The government is unable to protect them, not because of any discrimination but simply because its writ does not run in that part of the country. It is unable to protect men either. It may be true to say women would not fear attack but for the fact that they were women. *But I do not think that they would be regarded as subject to persecution within the meaning of the Convention. The necessary element of discrimination is lacking.*[67]

Although the assumption that the above scenario does not reflect discrimination can be challenged,[68] Lord Hoffman's express identification of discrimination as a 'necessary element' of 'persecution' is significant, and supports the argument advanced in this chapter. The approach of the NZIPT relegating the element of discrimination to a secondary position after it has been determined that the claimant has a well-founded fear of being persecuted diverges from the ordinary meaning of the notion of being persecuted within the context of the Refugee Convention and allows people who would not ordinarily be characterised as being persecuted to be so described. A better reading of Lord Hoffman is to see his distinction between 'persecution' and 'persecution which is based on discrimination' as reflecting the distinction identified in the range of definitions found in the *Oxford English Dictionary* and its French-language equivalent, discussed above, where the definition that comes closest to describing the ordinary meaning of the term for the purposes of the refugee definition is the one that incorporates the discriminatory element.

The interpretation is more faithful to the VCLT than the Hathaway definition because discrimination is inherent in the 'ordinary meaning', reinforced by reference to the immediate context of Article 1A(2) and the non-discrimination reference in the preamble, and further strengthened by the obligation to take into account relevant rules of international law applicable in the relations between the parties, namely the non-discrimination norm.

Indeed, understanding being persecuted with reference to the non-discrimination norm promotes systemic integration between international refugee law and international law more generally because it would preclude the reading in of a requirement to establish intention on the part of an actor of persecution (as suggested by pure dictionary-based definitions), since discrimination can be established irrespective of the intention of those

---

of human rights be 'serious'. It was argued that 'once a finding is made that there is a real chance of a breach of a human right which is of a sustained or systemic nature, this, without more, constitutes the requisite harm for the purposes of the "being persecuted" inquiry'. Counsel for the State in *DS (Iran)* argued against the interpretation of the being persecuted element advanced by Duffy J, arguing that 'serious harm' was integral to the notion – an argument with which the NZIPT agreed.

The NZIPT advances persuasive and uncontroversial reasons why serious harm is an integral aspect of being persecuted, and nothing in the proposed recalibration of dominant interpretations of the refugee definition advocated here seeks to displace that element from the analysis of the being persecuted element. It is therefore acknowledged that the approach taken by Duffy J in *CV v Immigration and Protection Tribunal* (n 65), whilst recognising discrimination to be an inherent feature of the experience of being persecuted, misses the mark by glossing over the requirement that any harm to which the claimant is exposed as part of her predicament be 'serious'.

[67] *Islam* (n 52) 654 (Lord Hoffman) (emphasis added).
[68] With thanks to Michelle Foster for discussion on this point.

responsible for discriminatory acts or omissions. In other words, and this is significant, the interpretation of being persecuted that has non-discrimination as a central organising principle adds to existing approaches in international law that focus on the predicament of the claimant, rather than the intentions of the actor of persecution, by systemically integrating the notion of being persecuted with principles of non-discrimination in IHRL, including in particular the principle that discrimination may be indirect or even systemic, as well as direct and intentional. The approach also remedies confusion around the point at which 'mere discrimination' becomes 'persecution'.

What, then, is to be made of the 'for reasons of' clause if it does not operate to bridge the act of persecution with the necessary element of discrimination reflected in the Convention grounds?

One approach would be to see the clause as narrowing the potential class of persons who may be recognised as Convention refugees by delimiting the categories of discrimination applicable. Thus, whereas Article 2 of the UDHR lists 'race, colour, sex, language, religion, political or other opinion, national or social origin, property, birth or other status' as prohibited grounds of discrimination, and Article 2 ICCPR also provides a non-exhaustive list affirming that the entitlements under the Covenant are to be enjoyed 'without distinction of any kind, such as race, colour, sex, language, religion, political or other opinion, national or social origin, property, birth or other status', Article 1A(2) of the Refugee Convention delimits the class of discriminated against persons entitled to recognition as refugees by including only those who face being persecuted 'for reasons of race, religion, nationality, membership of a particular social group or political opinion'.

This approach is supported by Lord Millett in *Islam*:

> Moreover, while the delegates in Geneva were willing to extend refugee status to the victims of discriminatory persecution, they were unwilling to define the grounds of persecution which would qualify for refugee status as widely as the discriminatory denial of human rights condemned by the Universal Declaration. Discriminatory persecution 'of any kind' would not suffice; the Convention grounds are defining, not merely illustrative as in the Universal Declaration.[69]

6.3.1 Challenges to the Proposition That Discrimination Is an Inherent Feature of Being Persecuted

In addition to the perspective articulated by the NZIPT in *DS (Iran)* set out above, the idea of discrimination being integral to the experience of being persecuted has previously been dismissed by Goodwin-Gill, who advances five arguments in response to the House of Lords judgment in *Islam*. The arguments are set out below before being treated in turn:

> Certainly, as both Lord Steyn and Lord Hoffmann noted, the Preamble to the 1951 Convention refers to UDHR48, but this was done with the intent to ensure that refugees' human rights were protected *after* they have fled their country of origin. It was not, and is not, the business of the 1951 Convention generally to promote non-discrimination or to protect the human rights of those who *might* become refugees. Other rules do this, and other mechanisms exist to make those rights effective.

---

[69] *Islam* (n 52) 660 (Lord Millett).

> ... Nothing in the *travaux préparatoires* suggests that the drafters had specifically in mind any notion of 'discriminatory denial of human rights', or equivalent formulation.
>
> ... That discourse lay in the future, and while it may be, and often is, possible to interpret persecution as some form of discriminatory denial of human rights, to think exclusively in these terms may fail to reflect the social reality of oppression. Approaching persecution as ineluctably linked to discrimination can work to advantage, of course, and has been adopted in various courts in various jurisdictions; but it remains a gloss on the original words, of which advocates need to be aware. A person who is persecuted by reason of a political opinion wrongly imputed in no way suffers discrimination by reason of some innate, unchangeable or otherwise fundamentally important characteristic. In that sense, he or she is 'innocent', 'untouched', 'unblemished', but still persecuted, either in error or deliberately, to set an example.[70]

Goodwin-Gill's first argument must be acknowledged. He argues that the non-discrimination reference in the preamble was concerned with the rights of refugees after recognition as such, rather than being reflective of a concern for their predicament in the country of origin. Cantor develops this challenge by recognising that the specific context referred to in the second perambular paragraph in which the UN had manifested its concern for refugees in the context of fundamental rights and freedoms concerned:

> 'discrimination practiced by certain States against immigrating labour and, in particular, against labour recruited from the ranks of refugees', and similar issues relating to the conduct of the host state entailing violations of the 'principle of non-discrimination embodied in the [UDHR]' by limiting refugees' 'access to facilities for accommodation, food, education, recreation and medical assistance'.[71]

Thus, basing the interpretation of being persecuted on the passing reference to non-discrimination in the second perambular paragraph is in itself inadequate. However, the first perambular paragraph is far broader and should be read with regard to the conventional language in the immediate context of Article 1A(2), where being persecuted 'for reasons of race, religion, nationality, membership of a particular social group or political opinion' guides attention to, in the words of Lord Hoffman, 'persecution which is based on discrimination',[72] and to have due regard to the non-discrimination norm as required by Article 31(3)(c).

Goodwin-Gill's second argument is rejected. He states that 'it was not, and is not, the business of the 1951 Convention generally to promote non-discrimination or to protect the human rights of those who *might* become refugees'. Recognising discrimination as being inherent to the experience of being persecuted for a Convention reason in no way requires states party to the Convention to 'promote' non-discrimination. The recognition simply requires that RSD proceed with a clear understanding of the constitutive elements of the refugee definition. Indeed, this book expressly rejects the proposition that RSD must be preoccupied with formal breaches of IHRL obligations.

---

[70] Guy S Goodwin-Gill, 'Judicial Reasoning and "Social Group" after *Islam* and *Shah*' (1999) 11 IJRL 538.
[71] Cantor, 'Defining Refugees' (n 36) 374, citing UNGA Resolution 315 (IV), 'Discrimination Practiced by Certain States against Immigrating Labour and, in Particular, against Labour Recruited from the Ranks of Refugees' (17 November 1949).
[72] *Islam* (n 52) 651 (Lord Hoffman).

Goodwin-Gill's third argument, namely that no reference is made to the discriminatory denial of human rights in the Convention preparatory works, has limited resonance. Beyond what appears to have been a passing observation that 'threat to freedom was a relative term and might not involve severe risks',[73] no discussion whatsoever appears to have taken place around the meaning to be attached to the being persecuted element.[74] This omission from the preparatory works of a central element of the refugee definition suggests that a similar omission of any reference to discrimination is unsurprising. Additionally, more general recourse to the divined intention of the parties needs to be balanced against the need for the treaty 'to be understood as presently speaking rather than forever defined by the circumstances in which it was conceived'.[75] As the importance and content of the non-discrimination obligation gains increasing clarity through doctrinal and jurisprudential development, and on the shoulders of the advances precipitated by the human-rights-based approach, its centrality to the predicament of the refugee in international law begins to emerge.

The fourth argument, namely that to focus on discrimination risks a failure 'to reflect the social reality of oppression', is addressed by the argument developed above that discrimination as a central organising principle invites, rather than discourages, consideration of both direct and indirect forms of conduct, as well as deeply entrenched structural or systemic discrimination. In other words, rather than deflecting attention, the approach developed in this book *focuses* attention on the discriminatory social context or 'the social reality of oppression'.

Finally, Goodwin-Gill's construction of a scenario in which a person is persecuted on account of an imputed political opinion misses the fact that there is no requirement under international law that a person actually possess the characteristic that exposes her to differential treatment.[76]

Thus, Goodwin-Gill's arguments against an interpretation of being persecuted that recognises discrimination as an inherent feature do not outweigh the arguments in favour of such an interpretation.

Aleinikoff adds:

> Furthermore, an anti-discrimination analysis may suggest additional norms that unduly restrict the scope of the Convention. It may lead adjudicators, for example, inappropriately to import into refugee law concepts from domestic anti-discrimination law, such as those relating to causation.[77]

Two observations address this point. First, as Foster notes, recourse to domestic anti-discrimination principles (as well as principles derived from IHRL) have supported broad, rather than narrow, approaches to RSD, in particular in relation to the question of causation.[78] To be sure, there are indeed examples of judicial authorities imposing

---

[73] Grahl-Madsen, *The Status of Refugees in International Law* (n 46) 193, citing UN Doc. E/AC.32/SR.20, 14 statement of Sir Leslie Brass (United Kingdom).
[74] Ibid   [75] James Hathaway, *The Rights of Refugees in International Law* (CUP 2005) 55.
[76] See discussion of general principles of non-discrimination in Section 6.1.
[77] T Alexander Aleinikoff, 'Protected Characteristics and Social Perceptions: An Analysis of the Meaning of "Membership of a Particular Social Group"' in Erika Feller, Volker Türk and Frances Nicholson (eds), *Refugee Protection in International Law: UNHCR's Global Consultations on International Protection* (CUP 2003) 292.
[78] '[T]he similarity between anti-discrimination law and refugee law has been noted by a number of common law courts, which have relied on flexible and liberal causation tests in the discrimination context as additional support for the adoption of a liberal standard in the refugee context', Foster,

additional requirements that purport to derive from non-discrimination principles. For example, the UK Court of Appeal in *Omoruyi* rejected a claim of a Christian man who, on account of his religious beliefs, refused to permit the ritual burial of his father as demanded by the Ogboni mafia, on the basis that discrimination 'is essential to the concept of persecution under the Convention ... [which requires] some element of *conscious* discrimination against the victim based on a Convention reason'.[79] The actors of persecution in this case were considered to be targeting him because of his conduct, as distinct from the religious belief underpinning that conduct.

Interestingly, counsel for the appellant had relied upon Goodwin-Gill's critique of the *Islam* judgment discussed above. However, Goodwin-Gill's argument that discrimination was not essential to the notion of being persecuted was rejected and the argument of counsel for the Secretary of State that 'discrimination is an essential feature of persecution for a Convention reason and that this requires the persecutor to be motivated by the reason in question, here religion' was accepted.[80] The clear difficulty with the judgment, however, is that in finding discrimination to be an essential feature of 'persecution for a Convention reason', the Court adopted a conception of discrimination that diverges from the accepted parameters under IHRL, not least the principle that a person may suffer from discrimination irrespective of the intentions of the relevant actors.[81] Critiquing the decision, Foster argues that:

> Reliance on anti-discrimination principles does not warrant the imposition of an intent test. On the contrary, as Madgwick J acknowledged in NACM, discrimination law 'treats as uncontroversial the proposition that discrimination may be legally established where either the intent or effect of conduct is discriminatory'.[82]

Thus, rather than disregard the ordinary meaning of being persecuted as set out above, the answer is to train adjudicators to apply international refugee law, and to have an appellate system in place that addresses errors of law.

6.4 Definition of Being Persecuted

Given that there is no controversy surrounding the proposition that discrimination has some role to play in defining the normative content of the legal status of being a refugee, does it matter whether its identification occurs as part of the being persecuted element as opposed to taking place when considering the nexus clause? It is submitted here that the

---

International Refugee Law and Socio-Economic Rights (n 51) 256, citing *Gafoor* (2000) 231 F 3d 645 (9th Cir. 2000), 671, *Chokov v Minister for Immigration and Multicultural Affairs* [1999] FCA 823, *R (on the application of Sivakumar) v Secretary of State for the Home Department* [2003] UKHL 14, [2003] 2 All ER 1097, [41].

[79] *Omoruyi* [2001] Imm AR 175, 181, cited in Foster, *International Refugee Law and Socio-Economic Rights* (n 51) 276 (emphasis added).

[80] Ibid   [81] See definitions and principles in Section 6.1.

[82] Foster, *International Refugee Law and Socio-Economic Rights* (n 51) 276, citing *NACM of 2002 v Minister for Immigration and Multicultural and Indigenous Affairs* [2003] FCA 1554 [59] (Madgwick J), as well as Lord Hoffman in *Sepet and Bulbul* [2003] 3 All ER 304, [2003] 1 WLR 856 who observed 'at any rate for the purposes of article 14 [of the European Convention] a law of general application may have a discriminatory effect if it contains no exceptions for people who have a right to be treated differently', at [30].

distinction is relevant, even if the full practical implications of this doctrinal turn remain to be seen.

Being persecuted entails more than serious harm + any kind of failure of state protection. In light of the foregoing, it is submitted here that being persecuted in the context of the Refugee Convention entails a condition of existence[83] in which discrimination[84] is a contributory cause of[85] (a real chance[86] of being exposed to) serious denials[87] of human rights[88] demonstrative of a failure of state protection. This definition would apply equally to political activists facing torture at the hands of state agents, to sexual and gender minorities facing a lifetime of dissimulation, and to members of marginalised ethnic or social groups facing differential exposure and vulnerability to serious disaster-related denials of the right to food, water, shelter, and the highest attainable standard of health. As noted in Chapter 5, the impact of this recalibrated definition is likely to be negligible in the vast majority of cases, and outcomes of decided cases would not likely differ if this recalibrated definition were to be substituted. The difference would most likely be detectable only in complex cases relating to the kind of systemic discrimination that engenders differential exposure and vulnerability to a range of denials of human rights where a single actor of persecution is not readily identifiable.

The purpose of proposing this recalibrated definition is not to change practice or to effect a change in outcomes in a particular case. The purpose is simply to highlight identified limitations in the dominant human-rights-based definition of being persecuted and to propose a recalibrated approach that is more faithful to the principles of treaty interpretation set out in Articles 31–33 VCLT.

Each element of this definition reflects arguments that have been developed across this book. First, it defines the experience of 'being persecuted', rather than 'persecution'. As Article 1A(2) itself requires a well-founded fear of 'being persecuted', the definition should reflect that predicament, rather than the conduct or intention of the actor(s) of persecution. The Hathaway definition that describes 'being persecuted' as a sustained or systemic violation of human rights demonstrative of a failure of state protection identifies the correct term to be defined, but goes on to describe conduct, rather than experience.

Second, by describing being persecuted as a condition of existence, the definition recognises the wider temporal scope of being persecuted, namely that a person will generally experience being persecuted before being exposed to specific acts of persecution. What distinguishes being persecuted from other forms of serious harm is the discriminatory context in which such harm is threatened or experienced. The enduring nature of the predicament, which the Hathaway definition recognises by reference to a sustained or

---

[83] A predicament-based approach as distinct from an event-based one. See discussion in Chapter 5.
[84] See discussion in this chapter.   [85] To the low contributory causation standard. See Section 3.2.
[86] Reflecting jurisprudence on risk assessment, including *Chan v Minister for Immigration and Ethnic Affairs* [1989] HCA 62, [1989] 169 CLR 379; *INS v Cardoza-Fonseca*, 480 US 421 [1987]; and *R v Secretary of State for the Home Department, ex parte Sivakumaran and Conjoined Appeals (UN High Commissioner for Refugees Intervening)* [1987] UKHL 1, [1988] AC 958, discussed in Chapter 5. Note also the possibility that a risk assessment may be superfluous to the determination as the conditions meeting the claimant on return may automatically entail a serious denial of human rights. See discussion in Chapter 5.
[87] Reflecting the rejection of the human-rights-adjudicatory approach described in Chapter 4.
[88] The harm element in human rights perspective. See discussion of the human-rights-based approach in Chapters 3.

systemic (risk), is an essential feature of being persecuted. This recognition of the broader temporal scope of the experience of being persecuted is also what enables consideration of systemic forms of discrimination (or 'structural violence') which is required if the full implications of placing the non-discrimination norm at the centre of the refugee definition are to be realised.

Third, the definition identifies the narrower personal scope, in which the experience of being discriminated against is inherent in the condition of being persecuted, rather than an afterthought to be considered subsequently to establishing that a person faces 'persecution'. Importantly, discrimination in this human-rights-based definition may be direct, indirect, or systemic, in line with established principles of international human rights law. The causal nexus between civil and political status and exposure to serious denials of human rights is signalled by the verb 'contributes to', reflecting the contributory cause approach to nexus described in Chapter 3.

Fourth, risk assessment, which, as Chapter 5 demonstrated, had migrated unhelpfully to the well-founded fear element of the definition, is reflected in the proposed new definition of being persecuted and retains the important insights from jurisprudence about the level of risk that can give rise to refugee status. Here, the distinction between this approach and other approaches is the express recognition that a person may face serious denials on return, *or* may face a condition of existence in which discrimination engenders a real chance of being exposed to such harm, and that both outcomes reflect the experience of being persecuted.

Fifth, the replacement of the term 'violations' with 'denials' reflects the rejection of the human-rights-adjudicatory approach discussed in Chapter 4. The requirement that such denials be 'serious' reflects the high threshold discussed, amongst other places, in *DS (Iran)*.

Finally, consistent with the surrogacy paradigm, a failure of state protection remains essential to the legal meaning of being persecuted under the Convention.

The role of state protection is integral to the Hathaway definition, and has supported the progressive development of international refugee law, not least in relation to gender-based persecution as reflected in *Islam*. Indeed, it was Lord Hoffman who embraced the pseudo-equation that persecution = serious harm + a failure of state protection. It was this bifurcated approach that brought violence within the 'private sphere' of the home into the 'public sphere' by recognising the possibility of the nexus to a Convention ground attaching to either the serious harm element or the failure of state protection element. At the same time, the notion of sufficiency of state protection has given rise to controversy in international refugee law, with debate around whether 'due diligence' or 'effective' standards should apply.

Apart from disagreements as to the formulation of a 'risk of being persecuted' set out in Chapter 5, it is not disputed that 'a decision on whether or not an individual faces a risk of "being persecuted" must also comprehend scrutiny of the state's ability and willingness effectively to respond to that risk'.[89] Indeed, a human-rights-based approach to interpreting the refugee definition would be very thin if it did not, as a matter of course, take into account searching questions about the ability and willingness of the state to fulfil its obligations under IHRL. The state is the primary duty-bearer under IHRL. States are also under a duty to address discrimination, both de jure and de facto, and when a person's civil

---

[89] Ibid 293.

or political status is a contributory cause of (a real chance of being exposed to) serious denials of human rights, there is a potential failure of state protection, irrespective of the efforts of the state to address the problem.

In light of the foregoing, it is contended that the definition proposed above is more faithful than the Hathaway definition to the principles of treaty interpretation set out at Articles 31–33 of the VCLT, whilst also reflecting key advances in international refugee law since the first version of that definition was advanced nearly thirty years ago. An avenue opens here for further research into the implications that adoption of this definition might have in other RSD contexts.

Understanding being persecuted as something related to, but distinct from, a specific event entailing a serious (although not necessarily discriminatory) denial of human rights helps to dismantle the event paradigm and thus guards against the dangerous tendencies towards reading an imminence requirement into international refugee law identified by Anderson, Foster, Lambert, and McAdam. The approach further invites a focus on structural violence, or systemic discrimination, that refugee law doctrine and jurisprudence has not developed extensively, potentially widening slightly the class of persons who are recognised as being eligible for refugee status. Such consideration is warranted given the need to interpret the refugee definition in line with the full scope of the non-discrimination norm as a relevant rule of international law applicable in the relations between the parties.

These are interesting avenues to pursue in further research. However, what remains for the purposes of this book is to consider how this recalibrated definition of being persecuted affects the determination of refugee status in the context of disasters and climate change.

7

Refugee Status Determination in the Context of 'Natural' Disasters and Climate Change

7.1 The Test

In light of the exercise in treaty interpretation performed in the preceding chapters, and following on from the recalibrated human-rights-based definition of being persecuted arrived at towards the end of Chapter 6, three questions will need to be asked in order to determine refugee status in any claim, whether related to disasters and climate change or not. This section sets out these questions, before addressing how each question may be approached in relation to claims for refugee status in the context of 'natural' disasters and climate change. The questions are:

1. Does the claim taken as a whole suggest that the claimant fears being persecuted for a Convention reason in her country of origin or former habitual residence, where that predicament is understood as a condition of existence in which discrimination on grounds of race, religion, nationality, membership of a particular social group, or political opinion is a contributing cause of (a real chance of being exposed to) serious denials of human rights demonstrative of a failure of state protection?
    a. If no, the person is not a refugee.
2. If yes, has the claim been established to the 'well-founded' standard?
    a. If no, the person is not a refugee.
3. If yes, is the person able to relocate internally?
    a. If yes, the person is not a refugee.
    b. If no, the person is a refugee.

Each of these three questions is considered in turn below.

7.2 Step One: Being Persecuted for a Convention Reason

The first step sets out to establish whether the claim, taken as a whole, suggests that the claimant fears being persecuted for a Convention reason in her country of origin or former habitual residence. It adopts the definition of being persecuted as a condition of existence in which discrimination is a contributing cause of (a real chance of being exposed to) serious denials of human rights demonstrative of a failure of state protection and integrates the Convention grounds in recognition of the unitary character of the refugee definition. Such integration is also practical as any consideration of whether a person faces discrimination will necessarily at that stage consider the grounds of discrimination. There are three elements in this first step:

- Does the claimant face discrimination on grounds of race, religion, nationality, membership of a particular social group, or political opinion?
- Does the claimant face (a real chance of being exposed to) serious denials of human rights?
- Is there a connection between discrimination and exposure to serious denials of human rights?

Each element is addressed in turn.

7.2.1 The First Element: Discrimination for a Convention Reason

The first element asks whether the claimant faces discrimination on grounds of race, religion, nationality, membership of a particular social group, or political opinion.

A profile must be constructed of the claimant in her social context, examining her unique position within the local and wider networks as well as the predicament of people similarly placed. Questions about where she lived, the kind of work she was engaged in, her health status, and so forth are relevant to the development of her individual profile. These issues are appropriately explored with reference to relevant human rights standards, for example relating to rights to adequate food and shelter, the right to water and the highest attainable standard of health as well as work-related rights under the ICESCR and of course the non-discrimination obligation.

Country of origin information will provide essential context against which this individual narrative falls to be considered, with familiar human rights reports being supplemented by material from human rights treaty monitoring bodies along with anthropological and other academic insights where available.

Having developed a sense of the claimant's status within her local and wider social context, a view can be reached about whether she is exposed to varying forms of discrimination, including direct, indirect, and systemic. Is this a person who experiences discrimination at all? What forms does that discrimination take, as manifestations of her relationship with both state and non-state actors? Relevant questions might relate to why she lives where she does and why she undertakes the kind of work she does. What role do state and non-state actors play in delimiting or enabling her options in this regard? In other contexts, this first element would elicit questions about whether people who express unwelcome political opinions or who belong to minority religious communities face discrimination, including being singled out for acts of violence, labour market discrimination, denial of access to education, and so forth.

If this first element is not satisfied, the person is not a refugee, as refugee status is inconceivable without the element of discrimination for a Convention reason, accurately understood from a human rights perspective.

7.2.2 The Second Element: (The Risk of) Exposure to Serious Denials of Human Rights

The second element asks whether the claimant faces (a real chance of being exposed to) serious denials of human rights.

This sub-question requires the decision-maker to consider the conditions awaiting the claimant on return. It invites consideration not only of the immediate conditions the claimant will face on arrival, but also of the risks associated with a foreseeable future disaster. In practice, a decision-maker ought not consider only the risks associated with disasters and climate change, but rather the full range of risks to which an individual claimant is exposed. Nonetheless, this subsection focuses on disaster risk given the focus of the book. Importantly, although reference is made to a 'foreseeable' future disaster, it is not suggested that a foreseeability standard be superimposed upon existing approaches to risk assessment in RSD. Foreseeability simply suggests that there is a recognisable hazard against which risk falls to be assessed.

As confirmed by the NZIPT in *AC (Tuvalu)*, past experience of being exposed to disaster-related harm raises the presumption that the claimant faces a risk of being exposed to a similar form of harm in the future, absent evidence of relevant disaster risk reduction measures taken by state and/or non-state actors.[1]

However, an important departure is called for from the standard articulated in *AF (Kiribati)* and reaffirmed in *AC (Tuvalu)* in relation to Articles 6 and 7 ICCPR, and the standard required under the Refugee Convention. When considering risk on return under the ICCPR, the NZIPT identifies a requirement that the risk assessment be focused on what the claimant may face within the immediate temporal scope of the return to her country of origin. This 'imminence' standard was derived from the admissibility criteria for bringing claims to the Human Rights Committee under the Optional Protocol, as articulated in the *Aaldersberg* case relating to nuclear weapons.[2] As Burson later acknowledged, there is no 'imminence' requirement in international refugee law. The question to ask instead is whether there is a real chance that the claimant will be exposed to a serious denial of human rights.

In the context of 'natural' disasters, two frames therefore invite consideration: risk on return during or in the immediate aftermath of a 'natural' disaster, and risk of being exposed to disaster-related denials of human rights in the context of a foreseeable future disaster. Risks to the enjoyment of the right to life, food, shelter, health as well as child and gender-specific rights are most likely to be in focus. In what follows, the two risk on return scenarios are considered in turn.

7.2.2.1 Ongoing Situations

Given the complexity of each 'natural' disaster situation, conventional distinctions between sudden-onset and slower-onset disasters should be avoided in the context of RSD. The focus must at all times be on the current circumstances and foreseeable developments, albeit placed in historical and social context.

Although much of the damage caused in cyclones, floods, earthquakes, and so forth may be repaired over weeks and months, individuals may nevertheless continue to experience serious denials of human rights even years after the impact.[3] Hence, no assumptions should be made that all risks have abated simply because the immediate aftermath has passed. A clear example of this reality is evident in the predicament reflected in the case of women in twenty-two camps for internally displaced persons in Port-au-Prince, Haiti, in which

---

[1] See discussion in *AC (Tuvalu)* [2014] NZIPT 800517-520 [69]. [2] See discussion in Section 3.7.
[3] See for example Kathleen Tierny and Anthony Oliver-Smith, 'Social Dimensions of Disaster Recovery' (2012) 30 IJMED 123.

high levels of exposure to gender-based violence persisted years after the 2010 Haitian earthquake.[4]

One difficulty in determining a real chance of exposure to serious denials of human rights on return during or in the aftermath of a disaster lies in obtaining sufficient facts about the unfolding conditions on the ground. Here, reports from the international humanitarian community are particularly relevant. The ReliefWeb website,[5] which is a specialised digital service of the UN Office for the Coordination of Humanitarian Affairs (UNOCHA), collates a vast array of reports from UN agencies as well as other organisations in a database enabling a range of filters relating to country, agency, and so forth. Situation reports, funding appeals, and similar documents can provide helpful insight into current risks. For example, more than one month after Typhoon Haiyan struck the Philippines, a 31 December 2013 UNOCHA situation report noted inter alia:

- Enormous shelter needs persist in affected areas. However, funding for the Shelter and Camp Coordination and Camp Management Clusters is lagging well behind overall funding levels.
- Funding of the shelter section of the Strategic Response Plan is currently at only 19.4 per cent. This is at odds with the huge shelter needs.
- The lack of good-quality CGI sheets continues to be a constraint.
- The Protection Cluster reports that partners have not provided NFIs or other assistance in Calubian, Leyte Province. To date, assistance received has been food from the Government.
- Municipalities in western Leyte still require WASH [Water, Sanitation and Hygiene] support.
- There is a lack of chlorination products for water purification in western Leyte.[6]

These kinds of reports thus point clearly to the conditions of existence the claimant would face on return.

Another helpful tool for both unfolding and foreseeable future disasters is the Integrated Phase Classification (IPC),[7] which assists in the assessment of the risk of being exposed to acute food insecurity. The tool is used to reflect the severity of a situation of acute food insecurity (which typically arises in the context of complex scenarios involving drought, conflict, and governance failures)[8] with reference to a number of indicators, including crude mortality rate, malnutrition prevalence, disease, food access/availability, dietary diversity, water access/availability, destitution/displacement, civil security, coping strategies, and livelihood assets. The scale contains five phases, with Phase 4 and 5[9] identified as emergency and famine, respectively:

---

[4] See IAHRC, PM 340/10 – Women and girls residing in 22 Camps for internally displaced persons in Port-au-Prince, Haiti <http://www.oas.org/en/iachr/decisions/precautionary.asp> accessed 16 April 2019.
[5] <http://www.reliefweb.int> accessed 15 April 2019.
[6] UNOCHA, 'Philippines: Typhoon Haiyan Situation Report No. 28 (as of 31 December 2013)' <https://www.undp.org/content/dam/philippines/docs/Typhoon%20Haiyan/OCHAPhilippinesTyphoonHaiyanN28_30December2013.pdf> accessed 15 April 2019.
[7] IPC Global Partners, 'Integrated Food Security Phase Classification Technical Manual Version 2.0: Evidence and Standards for Better Food Security Decisions' (Rome, FAO 2012) 20.
[8] See for example Stephen Devereaux, 'Famine in the Twentieth Century' (IDS Working Paper 105, 2000) <https://www.ids.ac.uk/files/dmfile/wp105.pdf> accessed 16 April 2019.
[9] Represented cartographically with the colour red and maroon respectively.

Phase 4 (Emergency): Even with any humanitarian assistance at least one in five HHs in the area have the following or worse: Large food consumption gaps resulting in very high acute malnutrition and excess mortality OR extreme loss of livelihood assets that will lead to food consumption gaps in the short term.

Phase 5 (Famine): Even with any humanitarian assistance at least one in five HHs in the area have an extreme lack of food and other basic needs where starvation, death, and destitution are evident. (Evidence for all three criteria of food consumption, wasting, and CDR[10] is required to classify Famine.)

Recognition by the international humanitarian community of a state of acute food insecurity is thus an indicator that an individual may be exposed to serious harm on return, and the IPC is thus a valuable practical tool for identifying situations in which hunger-related asylum claims could be explored. By way of example, Figure 7.1 illustrates Phases 4 and 5 food insecurity in Somalia in July 2011. Significantly, a declaration of acute food insecurity does not establish that the entire population faces a risk of starvation. Rather, for such a scale to be reached, only 20 percent of the households in the area need to be affected. The relevance of individual factors in this connection is therefore clear.

However, although acute food insecurity associated with drought is often characterised as being slow-onset, such situations change relatively rapidly with the intervention of humanitarian assistance (delivery of which can be complicated by, amongst other factors, armed conflict) and the onset of rains (provided they do not turn into floods). Thus an individualised assessment of the country conditions and the personal circumstances of the claimant will be necessary in each case. In the case of Somalia, for example, the food security situation improved steadily over the course of 2012, despite persistent pockets of emergency and crisis-level (Phase 3) insecurity. By February 2013, no areas of Somalia were reported as being in Phase 4 or 5 and the majority of territory was at Phase 2 (stressed).[11] The situation did not approach anywhere near 2011 conditions in the ensuing four years,[12] despite some predictions,[13] although the reality of changing donor funding priorities, combined with ongoing conflict, poor infrastructure, erratic climatic conditions, and other factors makes any recovery tenuous. Indeed, by 2017, the IPC indicated significant levels of acute food insecurity across the country, highlighting the recurrent nature of disaster risk. The (risk of) exposure to disaster-related harm in certain parts of the world is an enduring condition of existence. As emphasised in Chapter 2, this risk is differentially distributed, with marginalised groups disproportionately exposed and vulnerable.

7.2.2.2 Foreseeable Future Scenarios

The risk of being exposed to disaster-related harm does not return to zero once the relief teams depart (if they ever came in the first place). Rather, disaster risk is the dynamic outcome of changing patterns of hazard incidence, vulnerability, and exposure in a given social context. Although there is substantial discussion around calls to 'build back better', Paul notes that:

---

[10] CDR is the acronym for Crude Death Rate.
[11] FAO, 'Food Security and Nutrition Analysis Unit (FSNAU) – Somalia' <http://www.fsnau.org/ipc/ipcmap> accessed 16 April 2019.
[12] Ibid
[13] See for example the multi-agency 'Risk of Relapse: Call to Action: Somalia Country Update' (July 2014) <http://reliefweb.int/sites/reliefweb.int/files/resources/risk_of_relapse_-_call_to_action_-_somalia_crisis_update_july_-_2014_hi_res.pdf> accessed 16 April 2019.

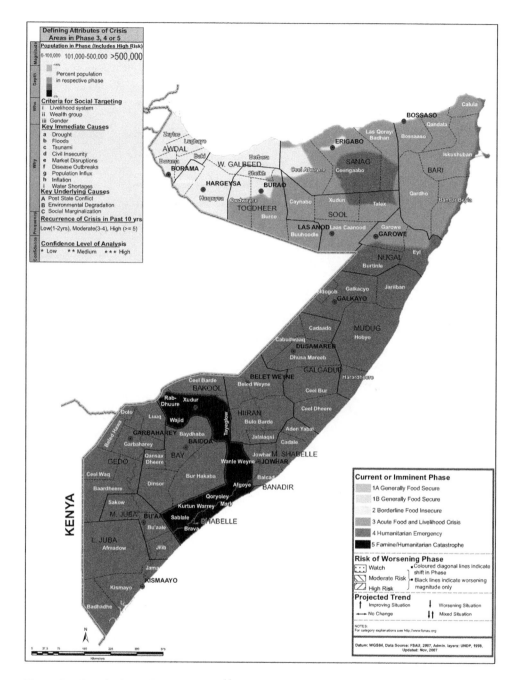

Figure 7.1 Somalia IPC July 2011 © FAO[14]

---

[14] FAO, 'Food Security and Nutrition Analysis Unit (FSNAU) – Somalia' <http://www.fsnau.org/ipc/ipc-map> accessed 16 April 2019. Reproduced with permission.

138   CLIMATE CHANGE, DISASTERS, AND THE REFUGEE CONVENTION

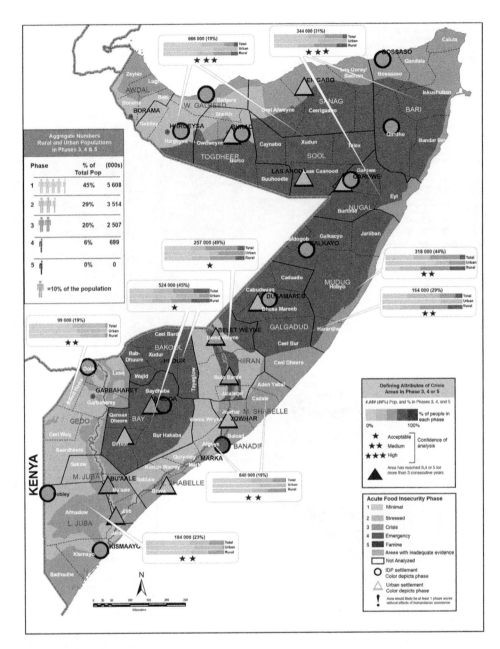

Figure 7.2 Somalia IPC June 2017 © FAO[15]

---

[15] FAO, 'FSNAU Food Security Quarterly Brief – Somalia' (April 2017) <http://www.fsnau.org/in-focus/fsnau-food-security-quarterly-brief-april-2017> accessed 16 April 2019. Reproduced with permission.

Relief activities typically include providing food, medicine, safe drinking water, sanitation services, temporary housing, security, as well as psychological and social support. Disaster relief is not designed to mitigate or reduce the risk of future disasters, but to return victims' level of living to a pre-disaster level. Relief workers are interested only in dealing with the immediate needs of disaster victims.[16]

Particularly for individuals living already marginal existences, a return to 'pre-disaster levels' entails a return to a situation of exposure and vulnerability, albeit potentially significantly more marginal than before the unfolding of the 'natural' disaster. Assessment of risk on return in the context of RSD should therefore not stop with consideration of the disaster relief cycle, but needs instead to examine foreseeable exposure to disaster-related harm in the context of a subsequent disaster.

Assessing risk in relation to a foreseeable future disaster entails a consideration of the likelihood that a particular hazard event or process will unfold in the foreseeable future, combined with reflection on the vulnerability and exposure of individuals and groups in particular social contexts. Although there is currently no single method for determining an individual disaster risk profile, an approach is outlined below that, it is argued, demonstrates the possibility of assessing risk in relation to a foreseeable future disaster.

Considering first the question of predicting when or whether a particular hazard event will unfold, the starting point is the recognition of two aspects of hazard prediction. First, one may be interested in determining *precisely* when the next cyclone may be expected to hit. Meteorologists are able to predict seasonal tropical cyclone patterns, and report their predictions in seasonal forecasts.[17] These forecasts, however, address regional activity and do not identify specific countries at risk. Targeted predictions are only possible a matter of days in advance of landfall. Even then, as the Australian Bureau of Meteorology notes, '[c]yclones vary considerably in their predictability. Some exhibit rapid changes in intensity or change course, speed up or slow down, primarily in response to changes in the surrounding environment'.[18] Thus, there is little to be gained from enquiring into the likely timing of the next cyclone. A similar conclusion can be reached in respect of earthquakes.[19] Floods, by contrast, tend to be more regular, although the timing of their occurrence is increasingly affected by anthropogenic factors.[20]

Instead of seeking precise predictions, the risk assessment in relation to a foreseeable future disaster may more appropriately focus on the overall likelihood that a hazard will unfold in the foreseeable future. Instead of looking to the future, this approach looks for patterns over

---

[16] Bimal Kanti Paul, *Environmental Hazards and Disasters: Contexts, Perspectives and Management* (1st edn, John Wiley & Sons 2011).

[17] See for example <http://www.tropicalstormrisk.com>, where forecasts are produced by scientists from University College London. Accessed 16 April 2019.

[18] <http://www.bom.gov.au/cyclone/about/accuracy.shtml> accessed 16 April 2019.

[19] On earthquakes, see for example Phil McKenna, 'Why Earthquakes Are Hard to Predict' *New Scientist* (14 March 2011) <https://www.newscientist.com/article/dn20243-why-earthquakes-are-hard-to-predict/> accessed 16 April 2019.

[20] See for example Rebecca Elmhirst, Carl Middleton and Bernadette P. Resurrección, 'Migration and Floods in Southeast Asia: A Mobile Political Ecology of Vulnerability, Resilience and Social Justice' in Carl Middleton, Rebecca Elmhirst and Supang Chantavanich (eds), *Living with Floods in a Mobile Southeast Asia: A Political Ecology of Vulnerability, Migration and Environmental Change* (Routledge 2018) who argue that 'policy decisions and their consequences – for example around urban growth, industrial and infrastructure development, deforestation and land and coastal degradation – contribute to the nature and frequency of floods', 4.

time in order to deduce the relative likelihood of a hazard event or process unfolding within a foreseeable timeframe. Thus, a simple search of the EM-DAT database reveals that the Philippines was hit by ninety-six tropical cyclones during 2005–15, making a rough annual average tropical cyclone incidence of 9.6.[21] A similar approach can be taken in relation to any other hazard. For example, Somalia has been affected by six droughts in ten years, with droughts recorded in five of the last seven years. Further insights may be gained by considering expert predictions of overall trends, including taking into account the possibility of climate change affecting the frequency and intensity of particular hazard events and processes. Individual narratives about past exposure to disasters also inform the risk profile.

The WorldRiskIndex, created by the Institute for Environment and Human Security at the United Nations University, attempts to determine the level of disaster risk facing a particular country by taking into consideration the country's exposure to particular hazards, as well as its ability to respond to such risks. The Index is designed to address four questions:

- How likely is an extreme natural event, and will it affect people?
- How vulnerable are people to natural hazards?
- To what extent can societies cope with acute disasters?
- Is a society taking preventive measures to face natural hazards to be reckoned with in the future?[22]

The Index considers risk in relation to the four natural hazards responsible for the greatest number of casualties and highest level of material damage between 1975–2005 (earthquakes, storms, floods, and droughts), plus the expected impact of sea level rise.

The approach expressly adopts the notion that disaster risk results from the interaction of a hazard event with exposed and vulnerable social conditions:

> In contrast to the assumption that a well-ordered society faces natural hazards and climate change, the concept of the WorldRiskIndex particularly underlines the importance of social, economic and environmental factors as well as governance aspects in determining whether a natural hazard will result in a disaster.[23]

The disaster risk score is a product of one score for a country's exposure to natural hazards and another for its vulnerability. Based on an evaluation of these indicators in 2016, societies where there is a high risk of being harmed in a 'natural' disaster include:

Madagascar	Haiti	Tanzania
Central African Republic	Zambia	Niger
Mozambique	Chad	Zimbabwe
Burundi	Eritrea	Togo
Liberia	Comoros	Sierra Leone[24]

---

[21] EM-DAT: The OFDA/CRED International Disaster Database, Université Catholique de Louvain, Brussels, Belgium <https://www.emdat.be> accessed 16 April 2019.

[22] Bündnis Entwicklung Hilft, *WorldRiskReport 2016: Focus: Logistics and Infrastructure* (Bündnis Entwicklung Hilft 2016) 10 <http://weltrisikobericht.de/wp-content/uploads/2016/08/WorldRiskReport2016.pdf> accessed 16 April 2019.

[23] Bündnis Entwicklung Hilft, *WorldRiskReport 2011: Focus: Governance and Civil Society* (Bündnis Entwicklung Hilft 2011) 14 <http://weltrisikobericht.de/wp-content/uploads/2016/08/WorldRiskReport_2011.pdf> accessed 16 April 2019.

[24] Bündnis Entwicklung Hilft, *WorldRiskReport* 2016 (n 22) 47.

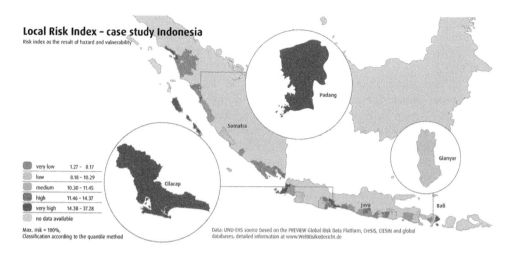

Figure 7.3 Local Risk Index: case study Indonesia © Bündnis Entwicklung Hilft 2011[25]

A similar approach to measuring risk of exposure to disaster-related harm is also possible at the local level. The 2011 WorldRiskReport employed the same methodology as used in the WorldRiskIndex to determine differential risk across Indonesia. The distribution of risk is made clear in Figure 7.3.

From this very local-level analysis, it is clear that disaster risk is not evenly distributed across the country. Rather, there are particular areas of very high vulnerability and exposure, such as Cilacap and Padang. The relevance of the internal relocation alternative is clearly evident in this connection, and will be revisited later in this chapter.

When individual risk factors, including vulnerability arising from past exposure to 'natural' disasters combined with marginal social position owing to discrimination, and other factors are considered against an overall risk outlook, it becomes possible to at least begin to consider incorporating the risk of being exposed to disaster-related harm in a future disaster into the determination of claims for refugee status. This is particularly the case in light of risk assessments based on tests such as a 'real chance', accompanied by the recognition that as little as a one in ten chance of being exposed to serious harm can suffice.[26] In this connection, it warrants repeating the observation of McHugh J in *Chan* to the effect that:

> The decisions in *Sivakumaran* and *Cardoza-Fonseca* also establish that a fear may be well-founded for the purpose of the Convention and Protocol even though persecution is unlikely to occur.[27]

Notwithstanding observations about the temporal scope of the experience of being persecuted set out in Chapter 5, the recognition by senior courts in three leading refugee law jurisdictions that international refugee law is highly capable of accommodating uncertainty is of paramount significance when considering claims relating to a risk of being exposed to

---

[25] Bündnis Entwicklung Hilft, WorldRiskReport 2011 (n 23) 41. Reproduced with permission.
[26] See discussion in Section 3.2.
[27] *Chan v Minister for Immigration and Ethnic Affairs* [1989] HCA 62, [1989] 169 CLR 379.

serious denials of human rights in the context of a foreseeable future disaster. Again, in any case where discrimination contributes to a risk of exposure to serious disaster-related harm, other forms of harm are also likely to manifest, and need to be considered with anxious scrutiny.

At the same time, there are differences between so-called sudden-onset disasters and slower-onset disasters, and establishing a well-founded fear of being persecuted for a Convention reason in relation to a foreseeable future disaster is perhaps more likely in the context of slower-onset hazard processes than where the hazard anticipated is a cyclone or an earthquake. As the preceding chapters have demonstrated, famine associated with drought is the scenario where, owing to the multiple social factors that make famine possible,[28] differential exposure and vulnerability is most readily identified. Famine is also predictable in ways that cyclones and earthquakes are not.

Evidence of risk in relation to foreseeable future acute food insecurity in more than thirty-five countries can be obtained from the Famine Early Warning Service (FEWS-NET). The methodology is described as a four-stage process, entailing building a knowledge base, monitoring existing conditions, forecasting, and classifying.

The analytical process is grounded in an in-depth understanding of factors that influence food security, such as markets and trade, agroclimatology, livelihoods, and nutrition. Insight into how poor households earn income, yearly averages for crop production and rainfall, and maps of regional trade flows are specific examples of the information available in this 'knowledge base'.[29]

Forecasting is built on the knowledge base and monitoring of conditions on the ground, informed by full-time analysts in twenty-two countries, and draws on a wider network of domestic and international expertise. The forecasts

> assess the current security situation in areas of concern, make assumptions about the future, and consider how those assumptions might affect food and income for poor households. Then, based on the convergence of evidence, they determine the most likely scenario and classify the expected level of food insecurity. Last, they identify major events or changes that could affect the outcome.[30]

Classification of risks is then performed in accordance with the Integrated Phase Classification as described above.

The result is a regional forecast of food insecurity where the risk of exposure to crisis, emergency, or famine-levels of food insecurity is clearly visualised and explained. For example, the food security forecast for February–May 2017 reflects crisis-level or higher food insecurity in parts of Somalia, Kenya, South Sudan, Ethiopia, Central African Republic, Nigeria, Democratic Republic of Congo, Zimbabwe, and Yemen.

The medium-term forecast for June–September 2017 reveals significant pockets of crisis-level or higher food insecurity in Somalia, Kenya, Ethiopia, Sudan, South Sudan, Chad, Nigeria, Tanzania, Democratic Republic of Congo, and Central African Republic, reflecting how, over time, risk not only changes but also endures. Even a passing awareness of ongoing conflicts in Yemen, South Sudan, Central African Republic, Somalia, Nigeria, and others will serve to reinforce understandings of the multiple factors that contribute to vulnerability and exposure to disaster-related harm.

---

[28] Consider Amartya Sen, *Poverty and Famines* (OUP 1981).
[29] <http://www.fews.net/our-work> accessed 16 April 2019.   [30] Ibid

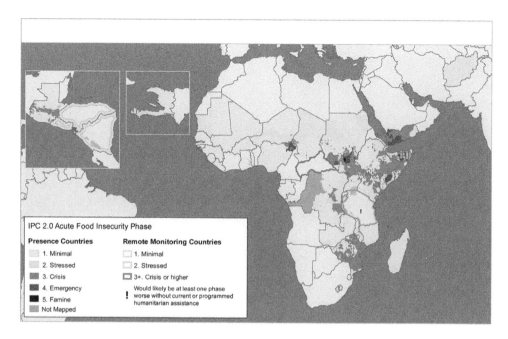

Figure 7.4 Acute food insecurity: near term (February–May 2017) © FEWS-NET[31]

It is perhaps a testament to the accuracy of the WorldRiskIndex that many of the countries identified as being 'high risk' in its 2016 report are represented in the Famine Early Warning System (FEWS) forecasts for 2017. Such confluence of assessments also suggests that it is possible to assess risk with some degree of accuracy in relation to foreseeable future disasters, even if much work remains to be done to refine an approach to assessing individual risk in this connection.

As climate change manifests in more frequent and protracted droughts, the likelihood that people in affected regions will be exposed to acute food insecurity is high, as a consequence of conflict over resources as well as restricted access to resources. It is equally likely that individuals who face discrimination in everyday life will be more exposed and vulnerable than people who are not so marginalised. Whether such people actually leave their home countries to seek recognition of refugee status in a state that has ratified the Refugee Convention is another matter.

If this second element is not satisfied, the person does not face being persecuted for a Convention reason and is therefore not a refugee.

7.2.3 The Third Element: Nexus

The third element asks whether there is a connection between discrimination and exposure to serious denials of human rights.

---

[31] <http://www.fews.net/fews-data/333> accessed 16 April 2019. Reproduced with permission.

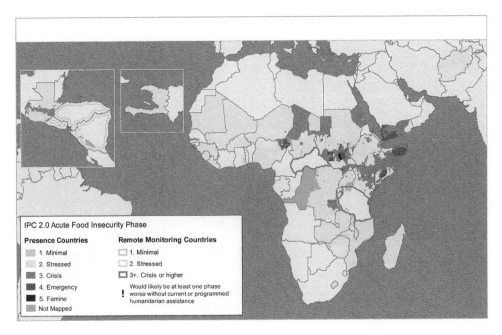

Figure 7.5 Acute food insecurity: medium term (June–September 2017) © FEWS-NET[32]

Importantly, the term 'contributory cause' in the recalibrated definition of being persecuted for a Convention reason recognises the need for a causal connection to be established between the discrimination and the real risk of exposure to serious denials of human rights. The question of causation, as explained in Chapter 3, has predominantly been addressed in relation to the 'nexus' or 'for reasons of' clause. Having surveyed the authorities, the conclusion was reached that, notwithstanding divergent domestic jurisprudence, there is some consensus in international refugee law on the question of the degree of causation that must be established between 'being persecuted' and the Convention grounds, and that any suggestion that a 'but for' or 'predominant cause' test is required should be rejected. Thus, for discrimination to be a contributory cause of (a real chance of being exposed to) serious denials of human rights, other factors can also contribute to that risk. What matters is that discrimination is not 'remote to the point of irrelevance'.[33]

Recalling insights from Chapter 2, decision-makers should have regard inter alia to both *ex ante* and *ex post facto* factors such as the location of shelters, the language of early warning and methods of communication, evacuation plans, the condition and location of housing, how relief is distributed, dynamics within camps/shelters, and so forth. Reflecting on the concept of structural violence, decision-makers should also consider the extent to which systemic disadvantage in relation to access to education, employment, shelter, food,

---

[32] <http://www.fews.net/fews-data/333> accessed 16 April 2019. Reproduced with permission.
[33] Michigan Guidelines on Nexus to a Convention Ground (2002) 23 Mich J Intl L 211. See discussion of causation in Section 3.2.

healthcare, and so forth restricts individual opportunities in a manner that is more than remotely constitutive of their exposure and vulnerability to disaster-related harm.

Although reports produced by humanitarian actors provide a good indication of the extent to which the minimum core of economic and social rights may be fulfilled, and also discuss other risk factors such as exposure to gender-based violence, they tend not to explore in detail the disaggregated risk along characteristics relevant to the Refugee Convention, necessitating recourse to the findings of more granular information from ethnographies and other academic research, NGO reports, and reports of UN human rights monitoring bodies as described above. An example can be provided in the context of the 2011 famine in Somalia.

According to the UNOCHA, by October 2011, as a result of immediate triggers of drought and crop failure, there were 750,000 people (17 percent of the regional population) living in famine conditions in South and Central Somalia.[34] Exploring the 'hidden dimensions of the Somali famine', Majid and McDowell identify the 'marginalized'[35] Reewin (Rahanweyn) and Bantu people living in that region as having been disproportionately affected.[36] Identifying what they consider to be the main cause of Reewin and Bantu vulnerability, the authors argue:

> Their main food, income and coping sources were strongly connected (their main sources of income and coping were derived by labouring on the farms which produced the food they purchased) creating high levels of vulnerability to a production shock, which in turn would diminish their principal source of coping.[37]

However, Reewin and Bantu people were not simply disproportionately exposed and vulnerable to famine because they happened to have insecure livelihoods. Rather, their predicament is inextricable from the wider social context. According to Cassanelli:

> the Bantu and Rahanweyn believe the major clans have systematically discriminated against them by excluding them from important government positions, restricting their educational opportunities and neglecting social and economic services in their home districts.[38]

Drawing on work by de Waal[39] and Hansch et al,[40] Majid and McDowell note that these two groups comprised the bulk of the 200,000–300,000 people who died during Somalia's last major famine in 1991–92.[41] Taking a historical perspective, the authors describe the marginalisation of more sedentary livelihood systems since the early 1960s, combined with targeted violence. Referring back to the 1991–92 famine, the authors explain that the Reewin and Bantu 'were the target of looting and violence by the more powerful major

---

[34] Nisar Majid and Stephen McDowell, 'Hidden Dimensions of the Somalia Famine' (2012) 1 Glob Food Secur 36, 36.
[35] Ibid [36] Ibid [37] Ibid 38.
[38] Lee Cassanelli, 'Victims and Vulnerable Groups in Southern Somalia' (Immigration and Refugee Board of Canada 1995) <http://www.refworld.org/docid/3ae6a8092.html> accessed 16 April 2019.
[39] Alex de Waal, 'Dangerous Precedents? Famine Relief in Somalia 1991–1993' in Joanna Macrae and Anthony Zwi (eds), *War and Hunger: Rethinking International Responses to Complex Emergencies* (Zed Books 1994).
[40] Steven Hansch et al, *Lives Lost, Lives Saved: Excess Mortality and the Impact of Health Interventions in the Somalia Emergency* (Refugee Policy Group 1994).
[41] Majid and McDowell, 'Hidden Dimensions of the Somalia Famine' (n 34) 37.

clan militias. Livestock and food stores were targeted'.[42] In relation to the Bantu, they explain that:

> The *Bantu* have a long history of forced removal from the most productive land by successive political authorities and exploitative relations with politically or militarily dominant groups. Large numbers became landless wage labourers under the Siad Barre regime.[43]

A 2004 report by the Danish Immigration Service explains:

> Somalia features a number of low status and minority groups which are frequently subject to abuse and exploitation. The Somali Bantu population is now the best known of these minorities representing about 5% of the total population. The Bantu are prone to theft of their land, rape, forced labour, and a range of discriminatory behaviour.[44]

In 'normal' times, the marginal experience of these groups has been held not to amount, in general, to being persecuted. In *DJ (Bantu – Not Generally at Risk) Somalia v Secretary of State for the Home Department*,[45] the UK Immigration and Asylum Tribunal, determined at paragraph 75 that:

> 1. Bantu generally are not a risk category. Bantu generally do not face a real risk of persecution or treatment contrary to Art 3 of the ECHR and they are able to obtain adequate protection from dominant Somali clans in their area.
> 2. Slavery or near-slavery is not the common lot of Bantu. Bantu generally do not face a real risk of treatment contrary to Art 4 of the ECHR.
> 3. It will only be in the unusual case that a Bantu will face a real risk of persecution or treatment contrary to Arts 3 or 4 of the ECHR on return. In assessing whether a case falls into the unusual category, it is important to take account of factors such as tribal clan history, geographical location, the nature of a person's previous relationship with the dominant Somali clan or clans in his or her area and the extent to which language, dialect and culture mark him or her out from surrounding clans and groupings.

Similarly, in *SH (Rahanweyn not a minority clan) Somalia*,[46] the Tribunal concluded that, although a finding by the lower instance Adjudicator that 'Members of the Rahanweyn clan have been persecuted in the past by majority clans' may be sustainable, the finding that 'the situation is still prevalent in Somalia' was not, owing to evidence that 'the Rahanweyn had established themselves through the RRA as the dominant clan in [the Bay of Bakool] area and it did not disclose any continuing power base held by the Hawiye in that area'.[47]

The far more detailed determination in *DJ* reveals a predicament where Bantu people live in a relationship of dependency on dominant clans that occupied the land they had traditionally farmed in situations described alternatively as 'sharecropping' and 'near-slavery':

---

[42] Ibid
[43] Ibid 38, citing Catherine Besteman and Lee Cassanelli (eds), *The Struggle for Land in Southern Somalia: The War Behind the War* (Westview Press/HAAN Publishing 2000).
[44] Cited in *AN (Tunni Torre)* [2004] UKIAT 270 (24 September 2004) [10].
[45] *DJ (Bantu - Not Generally at Risk) Somalia v Secretary of State for the Home Department CG* [2005] UKIAT 00089 (22 April 2005).
[46] *SH (Rahanweyn not a minority clan) Somalia CG* [2004] UKIAT 00272 (28 September 2004).
[47] Ibid [10–11].

According to the [2000 and 2002 Joint reports (see Appendix A)] conditions for Bantu reportedly vary according to the region in which they live ... As stated above Bantu have been largely displaced along the Juba and Shabelle rivers. They are usually able to remain in their home areas, to work mainly as labourers for the Somali clans (mainly the Marehan, Ogadeni and Habr Gedir) that have taken their traditional land. They can usually retain about 10% of their land for their own use ... However, in some cases Bantu work as plantation labourers in what Bantu elders describe as [a] situation of near slavery.[48]

People involved in such agricultural-based labour will, without adequate state or social safety nets, be extremely vulnerable to decreases in agricultural productivity in times of drought, such as when two consecutive rainy periods failed in Somalia between October 2010 and June 2011.[49] As sharecroppers and wage labourers in a relationship of dependency with dominant clans who have violently usurped land they had historically farmed, members of the Bantu ethnic group are, as a consequence of direct as well as systemic discrimination, differentially exposed and vulnerable to natural hazards. In this connection, Majid and McDowell argue:

> Clan and identity ... important in understanding long-term marginalization processes and outcomes, also help to understand the *differential vulnerability* to covariate risk, such as concurrent crop failure and related agricultural labor market collapse.[50]

The socio-economic marginalisation of these populations helps to explain, according to the authors, why the above 'proximate and complicating factors' caused famine 'predominantly in these communities and not others'.[51] Recalling the IPC, which identifies a situation of famine when only 20 percent of the population suffers from 'an extreme lack of food and other basic needs where starvation, death and destitution are evident',[52] the predicament of the disproportionately affected Bantu in the 2011 Somali famine reveals the discrimination behind such data.

Thus, *ex ante* historical and contemporary discrimination by state and non-state actors, and a lack of effective protection from a weak and at times failed state, caused or contributed to the differential vulnerability of Reewin and Bantu to the impacts of the drought-related famine, from which the state, weak as it was, was unable and/or unwilling to provide effective protection. Although, as case law relating to Reewin and Bantu asylum claims suggests, the challenges associated with their predicament would not always amount to 'persecution' outside of the famine context, as the disaster unfolded, a host of rights violations including the right to food, water, shelter, and the highest attainable standard of health became evident.[53] There is a prima facie case to be made that individuals belonging to Reewin and Bantu groups in Southern and Central Somalia would have been

---

[48] DJ (Bantu – Not Generally at Risk) (n 45) [16].
[49] Known locally as Deyr (October–December) and Gu (April–June 2011). IFRC, 'Emergency Appeal Operation Update: Somalia: Drought' (31 March 2012) <http://www.ifrc.org/docs/Appeals/12/MDRSO00106.pdf> accessed 16 April 2019.
[50] Majid and McDowell, 'Hidden Dimensions of the Somalia Famine' (n 34) 41 (emphasis added).
[51] Ibid 38. [52] IPC Global Partners, 'Integrated Food Security Phase Classification' (n 7) 20.
[53] Consider for example Francesco Checchi and W. Courtland Robinson, 'Mortality among Populations of Southern and Central Somalia Affected by Severe Food Insecurity and Famine during 2010–2012' (FAO/FEWSnet 2013) <http://www.fsnau.org/in-focus/study-report-mortality-among-populations-southern-and-central-somalia-affected-severe-food-> accessed 16 April 2019.

able to establish a well-founded fear of being persecuted for a Convention reason as soon as a real risk of exposure to serious denials of the right to food arose, and for at least the duration of the 2011 famine.[54] The FEWS data for the period after 2011 discussed above suggest that a real chance of being exposed to serious denials of human rights relating to acute food insecurity could not be established. The risk appeared to become real again only in 2017.

In the context of the 2011 famine, although other people also suffered from deprivation amounting to a denial of the minimum core of their right to food (amongst other potential violations), the predicament of Reewin and Bantu people is distinguishable because their particular vulnerability is closely connected to discrimination, which, together with the wider temporal scope of the predicament, is what distinguishes people who are generally exposed to serious denials of human rights from people who are persecuted in the same connection. It does not matter, on this approach, that no actors expressly singled out Reewin and Bantu to face serious harm, or even that any actor sought to exclude them from receiving disaster relief. It is sufficient that the evidence suggests that a person's civil or political status is a contributory cause of (a real chance of being exposed to) serious denials of human rights demonstrative of a failure of state protection.

Although the 2017 food insecurity was clearly foreseeable given the massive climatic as well as governance challenges Somalia faces, it will probably fall to case law to determine how far into the future decision-makers may look when assessing risk in relation to a foreseeable future disaster. Conflicts may end, economies may kick-start, investments may boom, and all of these may, potentially, affect an individual risk profile. There is no escaping the need for an individual assessment in each case, taking into account the humanitarian object and purpose of the Convention.

If this third element is not satisfied, then the person is not a refugee, as discrimination for a Convention reason must be established as a contributory cause of a (real chance of being exposed to) serious denials of human rights for such status to arise. If all three elements of step one are satisfied, then the decision-maker turns to the second step in the process.

7.3 Step Two: The Standard of Proof

The second step sets out to establish whether the claim has been made out to the 'well-founded' standard of proof.

Having argued in Chapter 5 that the function to be ascribed to the 'well-founded' criterion is to articulate a standard of proof that is applied to the conditions that are said to await the claimant, this second step entails an evaluation of the evidence to determine

---

[54] It warrants noting here that many will have been recognised as prima facie refugees in Kenya and Ethiopia under the 'expanded' definition under the Convention Governing the Specific Aspects of Refugee Problems in Africa (adopted 10 September 1969, entered into force 20 June 1974) 1001 UNTS 45. See Sanjula Weerasinghe, 'In Harm's Way: International Protection in the Context of Nexus Dynamics between Conflict or Violence and Disaster or Climate Change' (2018) UNHCR Legal and Protection Policy Research Series PPLA/2018/05 <https://www.unhcr.org/5c1ba88d4.pdf> accessed 16 April 2019. See also Tamara Wood, Technical Paper: Protection and Disasters in the Horn of Africa: Norms and Practice for Addressing Cross-Border Displacement in Disaster Contexts' (Nansen Initiative, January 2013) <http://www.nanseninitiative.org/wp-content/uploads/2015/03/190215_Technical_Paper_Tamara_Wood.pdf> accessed 16 April 2019.

whether, taken as a whole, the claim has been made out to a standard of proof that may be described as 'reasonable grounds'. This approach allows for limitations in available country of origin information and acknowledges the characterisation of RSD by Sedley J that:

> [A]djudication [of asylum claims] is not a conventional lawyer's exercise of applying a legal litmus test to ascertain facts; it is a global appraisal of an individual's past and prospective situation in a particular cultural, social, political and legal milieu, judged by a test which, though it has legal and linguistic limits, has a broad humanitarian purpose.[55]

If the person is unable to establish her claim to the 'well-founded' standard of proof, she is not a refugee. If her claim is established to the well-founded standard, then the decision-maker turns to the third and final step in the process.

7.4 Step Three: Internal Relocation

The third and final step asks whether the person would be able to relocate internally if returned to her country of origin or habitual residence.

Jurisprudence confirms that internal relocation may be precluded where the destination is affected by 'natural' disaster. In *Perampalam*, for example, the Federal Court of Australia explained that:

> It cannot be reasonable to expect a refugee to avoid persecution by moving into an area of grave danger, whether that danger arises from a natural disaster (for example, a volcanic eruption), a civil war or some other cause. A well-founded fear of persecution for a Convention reason having been shown, a refugee does not also have to show a Convention reason behind every difficulty for danger which makes some suggestion of relocation unreasonable.[56]

The spatial distribution of disaster risk raises obvious questions about the internal relocation alternative. Where a person may establish a well-founded fear of being persecuted for a Convention reason in Cilacap on Java, as revealed by the WorldRiskIndex above, cogent reasons would need to be adduced to support a contention that relocation to Ginyar on the same island would not be possible. At the same time, the IPC maps of Somalia during the 2011 famine show how widespread disaster risk can be.

People displaced internally in situations of disaster may be vulnerable to violence at the hands of both state and non-state agents, and also experience denials of economic and social rights. Consequently, even though the physical perimeter of a disaster situation may be delineated, an individual's risk profile may extend across the entire country. Reporting on the human rights situation of people internally displaced in the context of the 2004 Indian Ocean tsunami, the Special Representative of the Secretary General on Internal Displacement, Professor Walter Kälin, recounted:

> Protection concerns include access to assistance, discrimination in aid provision, enforced relocation, sexual and gender-based violence, recruitment of children into

---

[55] *R v IAT ex parte Shah* [1997] Imm AR 145, [153].
[56] *Perampalam v Minister for Immigration and Multicultural Affairs* [1999] FCA 165, [1999] 84 FCR 274, [19].

fighting forces, loss of documentation, safe and voluntary return or resettlement, and issues of property restitution.[57]

The experience of internal displacement in the context of disasters does not, in itself, reveal serious denials of human rights. However, such denials may arise on a case by case basis.

7.5 Circumstances Not Reflecting a Well-Founded Fear of Being Persecuted for a Convention Reason

The preceding sections have outlined relevant considerations for determining refugee status in the context of 'natural' disasters in a manner consistent with the social paradigm and the recalibrated approach to the refugee definition developed in this book. However, just as it is important to try to identify the kinds of circumstances in which a person may establish a well-founded fear of being persecuted in this connection, it is also important to clarify why a person may *not* satisfy the eligibility requirements at Article 1A(2) of the Refugee Convention.

When being persecuted is understood as a condition of existence in which discrimination is a contributory cause of (a real chance of being exposed to) serious denials of human rights demonstrative of a failure of state protection, the predicament of black residents in certain parts of New Orleans in the context of Hurricane Katrina in 2005 invites scrutiny. Likened by social scientists working within the social paradigm to the 1976 'class quake' in Guatemala,[58] research into the social causes of the 'natural' disaster and the social factors influencing the response revealed the extent to which discriminatory *ex ante* and *ex post* failures of state protection disproportionately adversely affected low-income black populations living in exposed areas.

In relation to *ex ante* failures of protection, the following have been noted:

- Systemic discrimination, including in relation to access to income generation, access to home loans contributing to segregation and black populations residing in more exposed areas[59]

---

[57] OHCHR, *Protection of Internally Displaced Persons in Situations of Disaster: A Working Visit to Asia by the Representative of the United Nations Secretary General on the Human Rights of Internally Displaced Persons Walter Kälin* (Office of the UN High Commissioner for Human Rights 2005) 8.

[58] See David Alexander, 'Symbolic and Practical Interpretations of the Hurricane Katrina Disaster in New Orleans', in *Understanding Katrina: Perspectives from the Social Sciences* <http://understandingkatrina.ssrc.org/Alexander/> accessed 16 April 2019, referring to the 'class-quake' discussed in Section 2.3.4.

[59] Kristin Henkel, John Dovidio and Samuel Gaertner, 'Institutional Discrimination, Individual Racism, and Hurricane Katrina' (2006) 6 Anal Soc Iss Pub Pol 99, 108. After explaining the factors that lead blacks to live in more exposed parts of town than whites, the authors explain: 'As a function of where they lived, when Hurricane Katrina hit, many Black people in New Orleans were already in a position to be disproportionately affected by the disaster. For example, HUD-funded public housing units above Feret Street West, which were occupied largely by Blacks, and New Orleans East were also on lower ground more vulnerable to flooding than higher, more desirable neighborhoods'. Similar observations are made in The Brookings Institution, 'New Orleans after the Storm: Lessons from the Past, a Plan for the Future' (Brookings Institution Metropolitan Policy Program, October 2005) 16 <http://www.brookings.edu/wp-content/uploads/2016/06/20051012_NewOrleans.pdf> accessed 16 April 2019. 'In this connection, sociology intersected quite exactly with geography and topography across the metropolitan area, and ensured that minority residents, poor people, and renters were all more likely to reside in the flood zone than white residents, better-off people, or those who owned their homes. The divides were sharp.'

- Failure to maintain the flood control system[60]
- Failure to have in place an early warning system that would reach low-income black and elderly populations[61]
- Failure to have an evacuation plan for the (predominantly low-income black and elderly) population that did not have access to their own transport[62]

Allegations of *ex post* failures of protection include:

- Discrimination in disaster relief[63]
- Indirectly discriminatory destruction of low-income housing[64]
- Indirectly discriminatory flood protection infrastructure reconstruction[65]

---

[60] The failure to maintain the flood control system was the subject of extensive civil litigation, in which the District Court of Louisiana found that the United States had negligently caused damage as a consequence of the failure to maintain the flood control system: See *In re Katrina Canal Breaches Consolidated Litigation (Robinson)* 647 F. Supp. 2d 644 (E.D. La. 2009) 733. 'As to the second inquiry, here it is manifestly evident that the Corps had a duty not to negligently expose the levee system along Reach II to harm, and it is likewise quite evident that if that levee system were harmed that there was great risk or harm to both people and property. In answer to the third question, such duty was obviously breached as extensively set forth in the findings of fact and conclusions of law set forth herein. Clearly, as to the fourth question, the risk of harm was within the scope of protection afforded by the duty breached as levees are designed to protect persons and property. The fifth question is likewise manifestly evident in that there were catastrophic damages that resulted from the breach. Therefore, this Court finds that the Corps of Engineers was negligent under the La. Civ. Code arts. 2315 and 2316 and is thus liable for damages arising from the destruction of the Reach 2 Levee.'

[61] Henkel, Dovidio and Gaertner, 'Institutional Discrimination, Individual Racism, and Hurricane Katrina' (n 59) 111–13, citing US House of Representatives, 'A Failure of Initiative: Final Report of the Select Bipartisan Committee to Investigate the Preparation for and Response to Hurricane Katrina' (15 February 2006) <https://www.gpo.gov/fdsys/pkg/CRPT-109hrpt377/pdf/CRPT-109hrpt377.pdf> accessed 16 April 2019.

[62] Ibid 106 and 108. Note that the authors emphasise the role of systemic discrimination and aversive racism, rather than direct and intentional formal or informal discrimination.

[63] See for example Henkel, Dovidio and Gaertner, 'Institutional Discrimination, Individual Racism, and Hurricane Katrina' (n 59) 109; Donald Saucier, Sara Smith and Jessica McManus, 'The Possible Role of Discrimination in the Rescue Response after Hurricane Katrina' (2007) *Journal of Race and Policy* 7, although note that this article describes itself as 'speculative' and focuses on applying theoretical models to general facts.

[64] See Davida Finger, 'Public Housing in New Orleans Post Katrina: The Struggle for Housing as a Human Right' (2011) 38 Rev Black Polit Econ 327, who inter alia cites the critique by the UN Human Rights Committee (Concluding Observations of the Human Rights Committee: United States of America, UN Doc. CCPR/C/USA/Q/3/CRP.4, para 22 (2006), the United Nations Special Rapporteur on Adequate Housing as a Component of the Right to an Adequate Standard of Living, and on the Right to Non-discrimination in this Context, along with the UN Independent Expert on Minority Issues (Office of the United Nations High Commissioner for Human Rights, 'UN Experts Call on United States to Protect African-Americans Affected by Hurricane Katrina' UN Doc. HR08023E (28 February 2008)) and UN Habitat Advisory Group on Forced Evictions (Advisory Group on Forced Evictions, An Advisory Group to UN Habitat, Mission Report to New Orleans, United States, Nov. 2010, at 31) of public housing allocation post-Katrina.

[65] Advocates for Environmental Human Rights (New Orleans, LA USA) and Peoples' Hurricane Relief Fund (New Orleans, LA USA), 'Hurricane Katrina: Racial Discrimination and Ethnic Cleansing in the United States in the Aftermath of Hurricane Katrina: A Response to the 2007 Periodic Report of the United States of America' (30 November 2007) 6 <www2.ohchr.org/english/bodies/cerd/docs/ngos/usa/USHRN23.doc> accessed 16 April 2019. Submissions to the Committee on the Elimination of Racial Discrimination asserted that improvement works would significantly disproportionately prioritise flood reduction measures in predominantly white neighbourhoods, with the consequence that 'African American residents are exposed to life-threatening conditions'.

As a point of departure, it is recognised that discrimination is widely considered to have played a role in both pre-disaster vulnerability and exposure to flooding, as well as in the response by authorities during and in the aftermath.[66] At the same time, it has also been judicially recognised that 'Hurricane Katrina was one of the most devastating hurricanes that has ever hit the United States, generating the largest storm surge elevations in the history of the United States'.[67] Hurricane Katrina therefore provides an excellent scenario through which to further explore the blurred boundary between human agency and 'natural' disasters, and the relevance of the distinction for RSD.

Two potential claims can be considered in this context. First, a person may have found herself overseas and felt compelled to apply for refugee status in order to avoid having to return in the aftermath of the 'natural' disaster. Second, a person may fear being exposed to serious denials of her human rights in the context of a foreseeable future disaster, and could seek recognition of refugee status on that basis. Might a person establish a well-founded fear of being persecuted for a Convention reason in either of these circumstances?

In relation to both scenarios, the decision-maker would need to gain insight into the profile of the individual in society. Are there indications that the individual faces discrimination? How does this discrimination manifest?

Focus would then shift to an assessment of the chance of exposure to serious denials of human rights. Here the approach differs depending on whether the claim relates to an ongoing 'natural' disaster and its aftermath or to a foreseeable future disaster.

For the claim relating to the ongoing 'natural' disaster, the decision-maker would need to obtain facts about unfolding conditions, which were readily available through media. A person returned to New Orleans in the aftermath of Hurricane Katrina would face serious difficulties accessing water, food, shelter, medical assistance, and may also face (a real chance of being exposed to) unlawful killing, sexual violence, and other forms of violent crime. The *Washington Post*, for example, reporting from the Ernest N. Morial Convention Center, decried the fact that:

> For five eternal-seeming days, as many as 20,000 people, most of them black, waited to be rescued, not just from the floodwaters of Hurricane Katrina but from the nightmarish place where they had sought refuge ... No one has been able to say how many people died inside the convention center; police, military and center officials estimate the number is about 10. Nor has there been any attempt to document the number of assaults, robberies and rapes that eyewitnesses said occurred from the time the first people broke into the convention center seeking shelter on the afternoon of Monday, Aug. 29, and when units of the Arkansas National Guard moved into the center on Friday, Sept. 2.[68]

However, despite very dramatic scenes reflecting (a real chance of being exposed to) serious denials of human rights in New Orleans in the immediate aftermath of the hurricane, state protection mechanisms began to address immediate threats to human rights in the ensuing

---

[66] See discussion below for sources. [67] *In re Katrina Canal Breaches* (n 60) 678.
[68] Wil Haygood and Ann Scott Tyson, 'It Was as if All of Us Were Already Pronounced Dead' *Washington Post* (15 September 2005) <http://www.washingtonpost.com/wp-dyn/content/article/2005/09/14/AR2005091402655.html> accessed 16 April 2019.

weeks, with the Federal Emergency Management Agency coordinating relief and as many as 1.3 million people displaced across all fifty states in the country.[69]

In light of these facts, and notwithstanding the need to carefully consider the individual risk profile as opposed to reaching conclusions based on the generality of information, it is unlikely that a decision-maker would find sufficient evidence that a person would, on return, face (a real chance of being exposed to) serious denials of human rights demonstrative of a failure of state protection. There would be little purpose to proceed to consider the relationship between discrimination and the individual (risk of) exposure to serious denials of human rights as such (risk of) exposure would be readily avoided by changing circumstances on the ground, including state protection, combined with an internal relocation alternative.

But what of the claimant who fears a repetition of Katrina, particularly in light of the apparent failures to 'build back better' in poorer, predominantly black, neighbourhoods? How might a decision-maker approach RSD in that scenario?

Adopting the approach set out in Step two of the test outlined above, the decision-maker would need to consider the likelihood of another 'natural' disaster unfolding. How foreseeable is it that New Orleans will be struck by another hurricane of a force sufficient to overwhelm reconstructed flood defences to the extent that a person would face (a real chance of being exposed to) serious denials of human rights demonstrative of a failure of state protection? Not only would meteorological data need to be considered, however; steps taken by the state (as well as non-state actors) to protect populations from foreseeable future hazard events would also have to be taken into account.[70] In the case of New Orleans, the evidence reveals that steps were being taken by the Army Corps of Engineers to 'build back better', even if the approach may well have reflected a prioritisation of certain parts of the city over others.

Even if it were established, in light of all of the evidence, that the claimant did indeed face (a real chance of being exposed to) serious denials of human rights in the context of a foreseeable future disaster, the decision-maker would need to consider the extent to which discrimination could be understood as being a contributory cause of this predicament. As the claimant already finds herself in another country, she would need to explain why she would find herself compelled to live in an exposed part of New Orleans, and what role discrimination had to do with that predicament, both in terms of her being compelled to reside in such an exposed area and in terms of the conditions giving rise to such exposure in the first place. It warrants recalling here that the standard of causation must necessarily be the 'contributory' standard described in Chapter 3, not a higher standard that would require discrimination to be the sole or predominant reason for exposure to harm.

In relation to her place of residence on return, the distinction between structure and agency referred to in Chapter 2 warrants recalling. An assumption might be that the claimant would be free to 'vote with her feet' and choose to live in an area that was less exposed to natural hazards. However, structural factors must also be taken into account, and in this connection it warrants recalling that in the case of Grenfell Tower, in which

---

[69] Christine Rushton, 'Timeline: Hurricane Katrina and the Aftermath' *USA Today* (24 August 2015) <https://www.usatoday.com/story/news/nation/2015/08/24/timeline-hurricane-katrina-and-aftermath/32003013/> accessed 16 April 2019.

[70] See *AC (Tuvalu)* (n 1) [69] discussed in Section 3.7.

many people died when fire engulfed the building in London,[71] was a social housing tower block,[72] meaning that individual residents would have had little choice but to live where they were allocated accommodation. The extent to which they were able to exercise agency, to choose to find better quality, safer accommodation, was circumscribed by their reliance on social assistance. Similar considerations would likely apply in relation to individuals fearing return to, for example, the exposed 9th Ward of New Orleans.

But the enquiry would need to look deeper still, as for a claimant to establish that discrimination is a contributory cause of (a real chance of being exposed to) serious denials of human rights demonstrative of a failure of state protection, questions would need to be asked about the reasons why the claimant would find herself amongst the category of persons in need of social assistance in the first place. The multiplicity of factors that will have influenced an individual trajectory in a wealthy country as large and diverse as the United States may potentially overwhelm an attempt to establish a sufficiently proximate connection between discrimination and the prospect of a claimant returning to a part of the country, let alone city, that is particularly exposed to natural hazards. Here, the observation by Joseph and Castan[73] about the difficulties associated with establishing an individual claim based on systemic discrimination resonates particularly clearly.

Evidentiary challenges would multiply when considering the role played by discrimination in the failure by the state to guarantee similar levels of flood protection across the city, as questions relating to decisions by federal, state, and local authorities in relation to urban planning, disaster risk reduction, economic development, social assistance, and so forth would need to be addressed. The scale of such an undertaking would potentially balloon and, absent a clear connection to discrimination, the claim would almost certainly be dismissed.

Of course, the availability of internal relocation would also reduce the prospects of success of such a claim to close to zero.

Thus, notwithstanding the disproportionate impact of Hurricane Katrina on low-income black and elderly populations in New Orleans, and the indications that in that connection discrimination was a contributory cause of (a real chance of being exposed to) serious denials of human rights demonstrative of a failure of state protection, an individual seeking recognition of refugee status in the aftermath of Katrina, or fearing exposure to a foreseeable future disaster, would be highly unlikely to establish that she had a well-founded fear of being persecuted for a Convention reason. However, each case would have to be assessed on its individual facts.

Thus, although it was established in the *Katrina Canal Breaches Consolidated Litigation* that the authorities had failed to fulfil certain disaster risk reduction obligations, and notwithstanding evidence of a disproportionate impact of the Katrina disaster on low-income black people living in the area, the individualised, forward-looking nature of RSD

---

[71] See discussion of budgetary choice not to ensure the building was fire-safe in Jonn Elledge, 'How the Fire at Grenfell Tower Exposed the Ugly Side of the Housing Boom' *New Statesman* (23 June 2017) <http://www.newstatesman.com/politics/uk/2017/06/how-fire-grenfell-tower-exposed-ugly-side-housing-boom> accessed 16 April 2019.

[72] See for example John Gapper, 'Grenfell: The Anatomy of a Housing Disaster' *Financial Times* (29 June 2017) <https://www.ft.com/content/5381b5d2-5c1c-11e7-9bc8-8055f264aa8b> accessed 16 April 2019.

[73] Sarah Joseph and Melissa Castan, *The International Covenant on Civil and Political Rights: Cases, Materials and Commentary* (3rd edn, OUP 2013) 818 discussed above in Section 6.1.

requires an evaluation of individual exposure and vulnerability to disaster risk, in light of the steps that have been taken by the state in the aftermath to reduce such risks.[74] Attempting to determine the role of discrimination in engendering disaster risk on the individual level will be particularly challenging in contexts where, despite evidence of systemic discrimination, the state does take steps to reduce disaster risk, and where structure does not overwhelm individual agency.

7.6 Conclusion

This book has addressed the application of the Refugee Convention in the context of disasters and climate change. Recognising epistemological as well as doctrinal limitations in existing judicial and academic engagement with the subject, a new perspective has been articulated. Epistemologically, this book has demonstrated why determination of refugee status in the context of disasters and climate change needs to take account of the wider social context in which disasters unfold, rather than being guided by outdated notions of the distinction between 'natural disasters' and human agency. Doctrinally, the book has emphasised the relevance of a human-rights-based interpretation of the refugee definition, but has proposed a recalibrated interpretation that sees discrimination as the central organising principle underpinning refugee status, and in this connection understands being persecuted as a condition of existence, as distinct from isolated 'acts of persecution'. This concluding chapter has demonstrated how addressing underlying epistemological and doctrinal assumptions can clarify how the Refugee Convention applies in the context of disasters and climate change.

Tackling the hazard paradigm that was shown to underpin the dominant view that the Refugee Convention does not apply in the context of disasters and climate change was largely a matter of consolidating decades worth of insights from the field of disaster risk reduction. Although this contribution is unique in terms of the express and detailed invocation of the 'social paradigm' as the appropriate lens for understanding what disasters are, the insights were readily accessible and applicable in the context of RSD.

More challenging has been the task of revisiting decades of doctrine and jurisprudence concerning the interpretation of the refugee definition. From the outset it has been clear that how the refugee definition is interpreted will have a significant impact on the determination of refugee status in the context of disasters and climate change, and Chapter 3 certainly confirmed this to be true.

Although the review of jurisprudence in Chapter 3 resulted in a rich taxonomy of the kinds of circumstances in which a person may establish a well-founded fear of being persecuted for a Convention reason in the context of disasters and climate change,[75] thereby already displacing the dominant view described in Chapter 1, it also highlighted underlying idiosyncrasies in established interpretations of the refugee definition. In particular, idiosyncrasies were identified relating to the temporal scope and the personal scope of being persecuted.

Applying the principles of treaty interpretation set out at Articles 31-33 VCLT resulted in a number of proposed recalibrations of the dominant human-rights-based interpretation

---

[74] As noted by the NZIPT in *AC (Tuvalu)* (n 1). See discussion in Section 3.7.
[75] Consolidated in tabular form in the Appendix.

of the refugee definition, culminating in a proposed definition of being persecuted as a condition of existence in which discrimination is a contributory cause of (a real chance of being exposed to) serious denials of human rights demonstrative of a failure of state protection. This new definition has been incorporated into a methodology for determining refugee status in the context of disasters and climate change that has been set out in this concluding chapter. Its application in other contexts giving rise to claims for recognition of refugee status remains to be examined, but it is suggested that the main practical impact will concern those cases in which failures of state protection result from systemic discrimination that is a contributory cause of a person's exposure to serious denials of human rights. Other, more direct forms of discrimination, are readily accommodated by the recalibrated definition, and outcomes are likely to be the same as when the dominant human-rights-based definition is employed.

The process described above has not resulted in an approach to RSD that promises to extend refugee status to the potentially very large numbers of people who may (and to some extent already do) cross international borders in the context of disasters and climate change. The refugee definition remains narrow. Indeed, one consequence of the process of interpretation conducted in this book is that the personal scope of being persecuted is narrowed from any class of person who faces a sustained or systemic denial of human rights demonstrative of a failure of state protection to the narrower class of person facing such a predicament as a consequence of her civil or political status. It now falls to legal practitioners to determine whether, on a case by case basis, such an approach contributes to the clear and principled determination of refugee status.

APPENDIX: TAXONOMY

Direct and international infliction of harm	Recognition of refugee status on the facts of the case	Recognition in principle	Suggested in doctrine or asserted in a claim for recognition of refugee status
Intentional environmental damage, as reflected in the predicament of the Marsh Arabs under Saddam Hussein	X	*AF (Kiribati)*	Kälin and Schrepfer (2012) McAdam (2012) King (2005) Burson (2010)
Crackdowns on (perceived) dissent relating to the causes and/or management of environmental degradation or disasters	*Refugee Appeal No. 76374* *RRT Case No. 0903555*	*AF (Kiribati)*	Kälin and Schrepfer (2012) *AL v Austria* *YC v Holder* *RRT Case No. 1001325* *RRT Case No. 060926579* *RRT Case No. 1104064*
Denial of disaster relief to members of opposition political parties, minority ethnic or religious groups, and so forth	*RN (Returnees) Zimbabwe CG* *Refugee Appeal No. 76237*	*Chan v Canada* *Hagi-Mohammed* *RS and Others (Zimbabwe - AIDS)* *AF (Kiribati)*	Kälin and Schrepfer (2012) McAdam (2012) *HS (returning asylum seekers) Zimbabwe CG*
Other failures of state protection	**Recognition of refugee status on the facts of the case**	**Recognition in principle**	**Suggested in doctrine or asserted in a claim for recognition of refugee status**
The state causes damage to the environment, or allows such conduct to be perpetrated by non-state actors, not caring about the adverse human impacts because of who the victims are	X	X	Kozoll (2004) Marcs (2008) *RRT Case N93/00894* (the 'Bangladesh cyclone case')

The state is unable to protect a population facing adversity in the context of a disaster	X	X	*Ferguson v Canada*
The state simply not 'being bothered' to protect a population facing adversity in the context of a disaster, or arbitrary refusal of international assistance for disaster relief	X	X	McAdam (2012) Hathaway (2014) *RRT Case No. 0907346* *Refugee Appeal No. 70965/98* *RRT Case No. 071295385* ('the Sri Lanka tsunami case') *RRT Case No. 1200203* ('the Australian Fukushima case')
Disaster risk management and response measures that amount to human rights violations for a Convention reason, such as in the context of forced relocation	X	X	Kälin and Schrepfer (2012)
Failure of disaster risk reduction and scenarios where state policies expose certain groups to disaster-related harm	X	*AF (Kiribati)* *AC (Tuvalu)*	McAdam (2012) Kälin and Schrepfer (2012)
Disasters engender serious threats to public order	Latin American cases referred to by Cantor (2015) and potentially Kenyan and Ethiopian cases referred to by Weerasinghe (2018) and Wood (2013)	*AC (Tuvalu)*	Kälin and Schrepfer (2012)
Ex ante discrimination is a contributory cause of (a real chance of being exposed to) serious denials of human rights demonstrative of a failure of state protection in circumstances where a person is exposed and vulnerable to disaster-related harm	X	X	This book

BIBLIOGRAPHY

Books and Monographs

Bankoff G, *Cultures of Disaster: Society and Natural Hazard in the Philippines* (RoutledgeCurzon 2003)
Cedervall Lauta K, *Disaster Law* (Routledge 2015)
Field C and others (eds), *Managing the Risks of Extreme Events and Disasters to Advance Climate Change Adaptation: Special Report of the Intergovernmental Panel on Climate Change* (CUP 2012)
Foster M, *International Refugee Law and Socio-Economic Rights: Refuge from Deprivation* (CUP 2007)
Goodwin-Gill GS and J McAdam, *The Refugee in International Law* (3rd edn, OUP 2007)
Grahl-Madsen A, *The Status of Refugees in International Law* (AW Sijthoff's Uitgeversmaatschappij NV 1966)
Guijt I and MK Shah, *The Myth of Community: Gender Issues in Participatory Development* (Intermediate Technology Publication 1998)
Hathaway J, *The Law of Refugee Status* (1st edn, Butterworths 1991)
 The Rights of Refugees in International Law (CUP 2005)
Hathaway J and M Foster, *The Law of Refugee Status* (2nd edn, CUP 2014)
Hewitt K (ed), *Interpretations of Calamity from the Viewpoint of Human Ecology* (Allen & Unwin Inc 1983)
Joseph S and M Castan, *The International Covenant on Civil and Political Rights: Cases, Materials and Commentary* (3rd edn, OUP 2013)
Kinnvall C and H Rydström (eds), *Climate Hazards, Disasters, and Gender Ramifications* (Routledge 2019)
Linderfalk U, *On the Interpretation of Treaties: The Modern International Law as Expressed in the 1969 Vienna Convention on the Law of Treaties* (Springer 2010)
McAdam J, *Climate Change, Forced Migration and International Law* (OUP 2012)
Middleton C, R Elmhirst and S Chantavanich (eds), *Living with Floods in a Mobile Southeast Asia: A Political Ecology of Vulnerability, Migration and Environmental Change* (Routledge 2018)
Nixon R, *Slow Violence and the Environmentalism of the Poor* (Harvard University Press 2013)
Noll G (ed), *Proof, Evidentiary Assessment and Credibility in Asylum Procedures* (Brill 2005)
O'Donnell M (ed), *Structure and Agency* (Sage 2010)
Paul BK, *Environmental Hazards and Disasters: Contexts, Perspectives and Management* (1st edn, John Wiley & Sons Ltd 2011)
Pobjoy J, *The Child in International Refugee Law* (CUP 2017)
Rodríguez H, E Quarantelli and R Dynes (eds), *Handbook of Disaster Research* (Springer 2006)
Schultz J, *The Internal Protection Alternative in Refugee Law: Treaty Basis and Scope of Application under the 1951 Convention Relating to the Status of Refugees and Its 1967 Protocol* (Brill 2018)
Sen A, *Poverty and Famines* (OUP 1981)
Vadenhole W, *Non-discrimination and Equality in the View of the UN Human Rights Treaty Bodies* (Intersentia 2005)
Vernant J, *The Refugee in the Post-War World* (George Allen & Unwin 1953)
Watts M, *Silent Violence: Food, Famine and Peasantry in Northern Nigeria* (University of California Press 1983)
Wisner B and others, *At Risk: Natural Hazards, People's Vulnerability and Disasters* (2nd edn, Routledge 2004)

Wisner B, JC Gaillard and I Kelman (eds), *The Routledge Handbook of Hazards and Disaster Risk Reduction* (Routledge 2012)

Zimmermann A (ed), *The 1951 Convention Relating to the Status of Refugees and Its 1967 Protocol: A Commentary* (OUP 2011)

Articles, Chapters, Reports and Occasional Papers

Advocates for Environmental Human Rights (New Orleans, LA USA) and Peoples' Hurricane Relief Fund (New Orleans, LA USA), 'Hurricane Katrina: Racial Discrimination and Ethnic Cleansing in the United States in the Aftermath of Hurricane Katrina: A Response to the 2007 Periodic Report of the United States of America' (30 November 2007) <www2.ohchr.org/english/bodies/cerd/docs/ngos/usa/USHRN23.doc> accessed 16 April 2019

Aleinikoff TA, 'Protected Characteristics and Social Perceptions: An Analysis of the Meaning of "Membership of a Particular Social Group"' in E Feller, V Türk and F Nicholson (eds), *Refugee Protection in International Law: UNHCR's Global Consultations on International Protection* (CUP 2003)

Alexander D, 'The Study of Natural Disasters, 1977–1997: Some Reflections on a Changing Field of Knowledge' (1997) 21 Disasters 284

― 'Symbolic and Practical Interpretations of the Hurricane Katrina Disaster in New Orleans' in *Understanding Katrina: Perspectives from the Social Sciences* <http://understandingkatrina.ssrc.org/Alexander/> accessed 17 April 2019

Anderson A and others, 'Imminence in Refugee and Human Rights Law: A Misplaced Notion for International Protection' (2019) 68 ICLQ 111

Barnett J, 'Security and Climate Change' (2003) 13 Global Environ Chang 7

Blake N, 'Luxembourg, Strasbourg and the National Court: The Emergence of a Country Guidance System for Refugee and Human Rights Protection' (2013) 25 IJRL 349

Bolin B, 'Race, Class, Ethnicity and Disaster Vulnerability' in H Rodríguez, EL Quarantelli and RR Dynes (eds), *Handbook of Disaster Research* (Springer 2007)

Briceño S, 'Forward' in B Wisner, JC Gaillard and I Kelman (eds), *The Routledge Handbook of Hazards and Disaster Risk Reduction* (Routledge 2012)

Brookings Institution, 'New Orleans after the Storm: Lessons from the Past, a Plan for the Future' Brookings Institution Metropolitan Policy Program (October 2005) <https://www.brookings.edu/wp-content/uploads/2016/06/20051012_NewOrleans.pdf> accessed 16 April 2019

Bündnis Entwicklung Hilft, 'WorldRiskReport 2011: Focus: Governance and Civil Society' (Bündnis Entwicklung Hilft 2011) <http://weltrisikobericht.de/wp-content/uploads/2016/08/WorldRiskReport_2011.pdf> accessed 16 April 2019

― 'WorldRiskReport 2016: Focus: Logistics and Infrastructure' (Bündnis Entwicklung Hilft 2016) <http://weltrisikobericht.de/wp-content/uploads/2016/08/WorldRiskReport2016.pdf> accessed 16 April 2019

Burson B, 'Environmentally Induced Displacement and the 1951 Refugee Convention: Pathways to Recognition' in T Afifi and J Jäger (eds), *Environment, Forced Migration and Social Vulnerability* (Springer 2010)

― 'Protecting the Rights of People Displaced by Climate Change: Global Issues and Regional Perspectives' in B Burson (ed), *Climate Change and Migration: South Pacific Perspectives* (Institute of Policy Studies 2010)

― 'The Concept of Time and the Assessment of Risk in Refugee Status Determination' (Presentation to Kaldor Centre Annual Conference, 18 November 2016) <http://www.kaldorcentre.unsw.edu.au/sites/default/files/B_Burson_2016_Kaldor_Centre_Annual_Conference.pdf> accessed 16 April 2019

Byrne R, 'James C. Hathaway and Michelle Foster, *The Law of Refugee Status*', Eur J Int Law, 26 (2015), 564

Cannon T, 'Reducing People's Vulnerability to Hazards: Communities and Resilience', (2008) United Nations University – WIDER Research Paper No. 2008/34 <https://www.wider.unu.edu/publication/reducing-people's-vulnerability-natural-hazards> accessed 4 April 2019

Cantor D, 'Background Paper: Law, Policy and Practice Concerning the Humanitarian Protection of Aliens on a Temporary Basis in the Context of Disasters' (States of the Regional Conference on Migration and Others in the Americas Regional Workshop on Temporary Protection Status and/or Humanitarian Visas in Situations of Disaster, San José, Costa Rica, 10–11 February 2015) <https://disasterdisplacement.org/wp-content/uploads/2015/07/150715_FINAL_BACKGROUND_PAPER_LATIN_AMERICA_screen.pdf> accessed 19 April 2019

'Defining Refugees: Persecution, Surrogacy and the Human Rights Paradigm'. In B Burson and D Cantor (eds.), *Human Rights and the Refugee Definition: Comparative Legal Practice and Theory* (Brill Nijhoff 2016)

Cassanelli L, 'Victims and Vulnerable Groups in Southern Somalia' (Immigration and Refugee Board of Canada 1995). <http://www.refworld.org/docid/3ae6a8092.html> accessed 16 April 2019

Checchi F and WC Robinson, 'Mortality among Populations of Southern and Central Somalia Affected by Severe Food Insecurity and Famine during 2010–2012', (FAO/FEWSnet 2013) <http://www.fsnau.org/in-focus/study-report-mortality-among-populations-southern-and-central-somalia-affected-severe-food-> accessed 16 April 2019

Chetail V, 'Are Refugee Rights Human Rights? An Unorthodox Questioning of the Relations between Refugee Law and Human Rights Law'. In R Rubio-Marin (ed.), *Human Rights and Immigration* (OUP 2014)

Cox TN, '"Well-Founded Fear of Being Persecuted": The Sources and Application of a Criterion of Refugee Status', Brooklyn J Int'l L, 10 (1984), 333.

Cutter SL, CT Emrich, JJ Webb and D Morath, 'Social Vulnerability to Climate Variability Hazards: A Review of the Literature', (2009) Final Report to Oxfam America <http://citeseerx.ist.psu.edu/viewdoc/download?doi=10.1.1.458.7614&rep=rep1&type=pdf> accessed 16 April 2019.

de Waal A, 'Dangerous Precedents? Famine Relief in Somalia 1991–1993'. In J Macrae and A Zwi (eds.), *War and Hunger: Rethinking International Responses to Complex Emergencies* (Zed Books 1994).

Devereaux S, 'Famine in the Twentieth Century', (IDS Working Paper 105, 2000) <https://www.ids.ac.uk/files/dmfile/wp105.pdf> accessed 16 April 2019.

Dörr O and K Schmalenbach, 'Article 31: General Rule of Interpretation'. In O Dörr and K Schmalenbach (eds.), *Vienna Convention on the Law of Treaties: A Commentary* (Springer 2012).

Durieux JF, 'Of War, Flows, Laws and Flaws: A Reply to Hugo Storey', Ref Survey Q, 31 (2012), 161.

Einarsen T, 'Drafting History of the 1951 Convention and the 1967 Protocol'. In A Zimmermann (ed.), *The 1951 Convention Relating to the Status of Refugees and its 1967 Protocol: A Commentary* (OUP 2011).

Elmhirst R, C Middleton and B Resurrección, 'Migration and Floods in Southeast Asia: A Mobile Political Ecology of Vulnerability, Resilience and Social Justice'. In C Middleton, R Elmhirst and S Chantavanich (eds.), *Living with Floods in a Mobile Southeast Asia: A Political Ecology of Vulnerability, Migration and Environmental Change* (Routledge 2018).

EM-DAT: The OFDA/CRED International Disaster Database, Université Catholique de Louvain, Brussels, Belgium <http://www.emdat.be> accessed 16 April 2019.

FAO, Food Security and Nutrition Analysis Unit (FSNAU), Somalia <http://www.fsnau.org/ipc/ipc-map> accessed 16 April 2019.

Farmer P, 'An Anthropology of Structural Violence', Curr Anthropology, 45(2004), 305.

Feller E, 'Statement by Ms. Erika Feller, Director, Department of International Protection, on the Refugee Definition' (Brussels, Strategic Committee for Immigration, Frontiers and Asylum (SCIFA), 6 November 2002) <http://www.refworld.org/docid/3dee02944.html> accessed 16 April 2019.

FEWS, Zimbabwe Food Security Outlook January–June 2010 <http://www.fews.net/sites/default/files/documents/reports/Zimbabwe_Outlook_January_2010_final.pdf> accessed 16 April 2019.

Zimbabwe Food Security Outlook October 2008–March 2009 <http://www.fews.net/sites/default/files/documents/reports/zimbabwe_outlook_2008_10.pdf> accessed 16 April 2019.

Finger D, 'Public Housing in New Orleans Post Katrina: The Struggle for Housing as a Human Right', Rev Black Polit Econ, 38 (2011), 327.

Fitzmaurice M, 'Interpretation of Human Rights Treaties'. In D Shelton (ed.), *The Oxford Handbook of International Human Rights Law* (OUP 2013).

Fletcher LE, E Stover and HM Weinstein, *After the Tsunami: Human Rights of Vulnerable Populations* (Human Rights Center of UC Berkeley and East-West Center 2005).

Gaillard JC, 'Caste, Ethnicity, Religious Affiliation and Disaster'. In B Wisner, JC Gaillard and I Kelman (eds.), *The Routledge Handbook of Hazards and Disaster Risk Reduction* (Routledge 2012).

Galtung J, 'Violence, Peace, and Peace Research', J Peace Res, 6 (1969), 167.

Gill T, 'Making Things Worse: How "Caste-Blindness" in Indian Post-Tsunami Disaster Recovery Has Exacerbated Vulnerability and Exclusion', (2007) Dalit Network Netherlands <http://idsn.org/uploads/media/Making_Things_Worse_report.pdf> accessed 16 April 2019.

Gleick P, 'Global Climatic Change and International Security', Colo J Int'l Envtl L & Pol'y, 1 (1990), 41.

'Water, Drought, Climate Change, and Conflict in Syria', Weather Clim Soc, 6 (2014), 331.

Goodwin-Gill GS, 'Judicial Reasoning and "Social Group" after *Islam* and *Shah*', IJRL, 11 (1999), 538.

'Unaccompanied Refugee Minors: The Role and Place of International Law in the Pursuit of Durable Solutions', Int'l J Child Rts, 3 (1995), 405.

Hansch S, S Lillibridge, G Egeland, C Teller and M Toole, *Lives Lost, Lives Saved: Excess Mortality and the Impact of Health Interventions in the Somalia Emergency* (Refugee Policy Group 1994).

Hathaway J, 'Food Deprivation: A Basis for Refugee Status?', Soc Res, 81 (2014), 327 <http://repository.law.umich.edu/articles/1076/> accessed 4 April 2019.

Hathaway J and M Foster, 'Internal Protection/Relocation/Flight Alternative as an Aspect of Refugee Status Determination'. In E Feller, V Türk and F Nicholson (eds.), *Refugee Protection in International Law: UNHCR's Global Consultations on International Protection* (CUP 2003).

'The Causal Connection ("Nexus") to a Convention Ground, Discussion Paper No. 3, Advanced Refugee Law Workshop International Association of Refugee Law Judges Auckland, New Zealand, October 2002' IJRL, 15 (2003), 461.

Hathaway J and W Hicks, 'Is There a Subjective Element in the Refugee Convention's Requirement of "Well-Founded Fear"?', Mich J Int'l L, 26 (2005), 505.

Hathaway J and J Pobjoy, 'Queer Cases Make Bad Law', NYU J Int'l L & Pol, 44 (2011), 315.

Henkel K, J Dovidio and S Gaertner, 'Institutional Discrimination, Individual Racism, and Hurricane Katrina', Anal Soc Iss Pub Pol, 6 (2006), 99.

Hewitt K, 'The Idea of Calamity in a Technocratic Age'. In K Hewitt (ed.), *Interpretations of Calamity from the Viewpoint of Human Ecology* (Allen and Unwin Inc 1983).

'Excluded Perspectives in the Social Construction of Disaster', IJMED, 13 (1995), 317.

Hilhorst D, 'Unlocking Disaster Paradigms: An Actor-Oriented Focus on Disaster Response' (Abstract Submitted for Session 3 of the Disaster Research and Social Crisis Network Panels of the 6th European Sociological Conference, 23–26 September, Murcia, Spain, 2003) available on request from https://www.researchgate.net/publication/254033796_Unlocking_disaster_paradigms_An_actor-oriented_focus_on_disaster_response> accessed 16 April 2019.

Houghton R, T Wilson, W Smith and D Johnston, '"If There Was a Dire Emergency, We Never Would Have Been Able to Get in There": Domestic Violence Reporting and Disasters', IJMED, 28 (2010), 270.

IARLJ-Europe, 'Qualification for International Protection (Directive 2011/95/EU): A Judicial Analysis', (EASO 2016) <https://www.easo.europa.eu/sites/default/files/QIP%20-%20JA.pdf> accessed 16 April 2019.

IFRC, 'World Disasters Report 2007: Focus on Discrimination', (2007) <http://www.ifrc.org/docs/Appeals/12/MDRSO00106.pdf> accessed 16 April 2019.

'Emergency Appeal Operation Update: Somalia: Drought', (31 March 2012).

What Is a Disaster? <https://www.ifrc.org/en/what-we-do/disaster-management/about-disasters/what-is-a-disaster/> accessed 16 April 2019.

Jaeger G, 'On the History of the International Protection of Refugees', Int'l Rev Red Cross, 83 (2001), 727.

Kälin W and N Schrepfer, 'Protecting People Crossing Borders in the Context of Climate Change: Normative Gaps and Possible Approaches', (2012) UNHCR Legal and Protection Policy Research Series PPLA/2012/01 <http://www.unhcr.org/4f33f1729.pdf> accessed 16 April 2019.

King T, 'Environmental Displacement: Coordinating Efforts to Find Solutions', Geo Int'l Envtl L Rev, 18 (2005), 543.

Kolmannskog V, 'Climate Change, Environmental Displacement and International Law', J Int Dev, 24 (2012), 1071.

Kozoll C, 'Poisoning the Well: Persecution, the Environment, and Refugee Status', Colo J Int'l Envtl L & Pol'y, 15 (2004), 271.

Lambert H, 'The Conceptualisation of "Persecution" by the House of Lords: *Horvath vs Secretary of State*', IJRL, 13 (2001), 16, 28.

Lillich RB, 'Civil Rights'. In T Meron (ed.) *Human Rights in International Law: Legal and Policy Issues* (OUP 1984).

Mackey A, M Treadwell, B Dingle and B Burson, 'A Structured Approach to the Decision Making Process in Refugee and other International Protection Claims' (Presented at the IARLJ/JRTI/UNHCR conference 'The Role of the Judiciary in Asylum and Other International Protection Law in Asia', Seoul, Korea, 10–11 June 2016) [5] <https://www.iarlj.org/iarlj-documents/general/IARLJ_Guidance_RSD_paper_and_chart.pdf> accessed 16 April 2019.

Macklin A, 'The Law of Refugee Status 2d ed.' Hum Rts Q, 39 (2017), 220.

Majid M and S McDowell, 'Hidden Dimensions of the Somalia Famine', Glob Flood Sec, 1 (2012), 36.

Marcs C, 'Spoiling Movi's River: Towards Recognition of Persecutory Environmental Harm within the Meaning of the Refugee Convention', Am U Int'l L Rev, 24 (2008), 31.

McAdam J, 'Seeking Asylum under the Convention on the Rights of the Child: A Case for Complementary Protection', Int'l J Child Rts, 14 (2006), 251.

'Interpretation of the 1951 Convention'. In A Zimmermann (ed.), *The 1951 Convention Relating to the Status of Refugees and Its 1967 Protocol: A Commentary* (OUP 2011).

McKenna P, 'Why Earthquakes Are Hard to Predict', *New Scientist* (14 March 2011) <https://www.newscientist.com/article/dn20243-why-earthquakes-are-hard-to-predict/> accessed 16 April 2019.

Michigan Guidelines on Nexus to a Convention Ground, Mich J Int'l L, 23 (2002), 211.

Musalo K, 'Irreconcilable Differences? Divorcing Refugee Protections from Human Rights Norms', Mich J Int'l L, 15 (1993), 1179.

Nansen Initiative, 'Agenda for the Protection of Cross-Border Displaced Persons in the Context of Disasters and Climate Change', (2015) <https://disasterdisplacement.org/the-platform/our-response> accessed 4 April 2019.

Ní Ghráinne B, 'The Internal Protection Alternative Inquiry and Human Rights Considerations – Irrelevant or Indispensable?', IJRL, 27 (2015), 29.

Noll G, 'Evidentiary Assessment under the Refugee Convention: Risk, Pain and the Intersubjectivity of Fear'. In G Noll (ed.), *Proof, Evidentiary Assessment and Credibility in Asylum Procedures* (Brill 2005).

Norwegian Refugee Council/Internal Displacement Monitoring Centre (NRC/IDMC), 'The Nansen Conference: Climate Change and Displacement in the 21st Century' (7 June 2011) <http://www.refworld.org/docid/521485ef4.html> accessed 16 April 2019.

O'Keefe P, K Westgate and B Wisner, 'Taking the Naturalness out of Natural Disasters', Nature, 260 (1976), 566.

Oliver-Smith A, 'Anthropological Research on Hazards and Disasters', Annu Rev Anthropol, 25 (1996), 303.

'"What Is a Disaster?" Anthropological Perspective on a Persistent Question'. In A Oliver-Smith and S Hoffman (eds.), *The Angry Earth: Disaster in Anthropological Perspective* (Psychology Press 1999).

'Haiti's 500-Year Earthquake'. In M Schuller and P Morales (eds.), *Tectonic Shifts: Haiti Since the Earthquake* (Kumarian Press 2012).

Peacock WG with AK Ragsdale, 'Social Systems, Ecological Networks and Disasters: Toward a Socio-Political Ecology of Disasters'. In WG Peacock, BH Morrow and H Gladwin (eds.), *Hurricane Andrew: Ethnicity, Gender and the Sociology of Disasters* (Routledge 1997).

Perry R, 'What Is a Disaster?'. In H Rodríguez, E Quarantelli and R Dynes (eds.), *Handbook of Disaster Research* (Springer 2007).

Rodríguez H and J Barnshaw, 'The Social Construction of Disasters: From Heat Waves to Worst-Case Scenarios', Contemp Sociol, 35 (2006), 218.

Saucier D, S Smith and J McManus, 'The Possible Role of Discrimination in the Rescue Response After Hurricane Katrina', J Race Pol, 3, (2007), 113–121.

Scott M, 'Finding Agency in Adversity: Applying the Refugee Convention in the Context of "Natural" Disasters and Climate Change', Ref Survey Q, 35 (2016), 26.

Serje J, 'Data Sources on Hazards'. In B Wisner, JC Gaillard and I Kelman (eds.), *The Routledge Handbook of Hazards and Disaster Risk Reduction* (Routledge 2012).

Shacknove A, 'Who Is a Refugee?', Ethics, 95 (1985), 274.

Söderbergh C, 'Human Rights in a Warmer World: The Case of Climate Change Displacement', (2011) Working Paper 2011-01-28 <http://lup.lub.lu.se/record/1774900> accessed 16 April 2019.
Storey H, 'Persecution: Towards a Working Definition' (2014). In V Chetail and C Bauloz (eds.), *Research Handbook on International Law and Migration* (Edward Elgar 2014).
  'What Constitutes Persecution? Towards a Working Definition', IJRL, 26 (2014), 272.
  'The Law of Refugee Status, 2nd edition: Paradigm Lost?', IJRL, 27 (2015), 348.
Tierny K and A Oliver-Smith, 'Social Dimensions of Disaster Recovery', IJMED, 30 (2012), 123.
Twigg J, 'Disaster Risk Reduction: Mitigation and Preparedness in Development and Emergency Programming', ODI Humanitarian Practice Network Good Practice Review No. 9 (March 2004) <https://www.preventionweb.net/educational/view/8450> accessed 4 April 2019.
US House of Representatives, 'A Failure of Initiative: Final Report of the Select Bipartisan Committee to Investigate the Preparation for and Response to Hurricane Katrina', (15 February 2006) <https://www.gpo.gov/fdsys/pkg/CRPT-109hrpt377/pdf/CRPT-109hrpt377.pdf> accessed 16 April 2019.
Weerasinghe S, 'In Harm's Way: International Protection in the Context of Nexus Dynamics between Conflict or Violence and Disaster or Climate Change', (2018) UNHCR Legal and Protection Policy Research Series PPLA/2018/05 <https://www.unhcr.org/5c1ba88d4.pdf> accessed 16 April 2019.
Wisner B, JC Gaillard and I Kelman, 'Framing Disaster: Theories and Stories Seeking to Understand Hazards, Vulnerability and Risk'. In B Wisner, JC Gaillard and I Kelman (eds.), *The Routledge Handbook of Hazards and Disaster Risk Reduction* (Routledge 2012).
Wood T, 'Technical Paper: Protection and Disasters in the Horn of Africa: Norms and Practice for Addressing Cross-Border Displacement in Disaster Contexts', (Nansen Initiative, January 2013) <http://www.nanseninitiative.org/wp-content/uploads/2015/03/190215_Technical_Paper_Tamara_Wood.pdf> accessed 16 April 2019.
Zimmermann A and C Mahler, 'Article 1A, para. 2 (Definition of the Term "Refugee"/Définition du Terme "Réfugié")'. In A Zimmermann (ed.), *The 1951 Convention Relating to the Status of Refugees and Its 1967 Protocol: A Commentary* (OUP 2011).

United Nations Documents

CERD, 'Concluding Observations of the Committee on the Elimination of Racial Discrimination: Brazil', (28 April 2004) CERD/C/64/CO/2.
CESCR, 'General Comment No. 3: The Nature of States Parties' Obligations (Art. 2, Para. 1, of the Covenant)', (14 December 1990) E/1991/23.
  'General Comment No. 12 on the Right to Adequate Food', (12 May 1999) E/C.12/1999/5.
  'General Comment No. 20: Non-discrimination in Economic, Social and Cultural Rights (Art. 2, Para. 2, of the International Covenant on Economic, Social and Cultural Rights)', (2 July 2009) E/C.12/GC/20.
Conference of Plenipotentiaries on the Status of Refugees and Stateless Persons, Summary Record of the Twenty-Second Meeting, (26 November 1951) UN Doc A/CONF.2/SR.22.
CRC, 'General Comment No. 13 (2011): The Right of the Child to Freedom from All Forms of Violence', (18 April 2011) CRC/C/GC/13.
CRPD, 'General Comment No. 3: Article 6: Women and Girls with Disabilities', (2 September 2016) CRPD/C/GC/3.
HRC, 'General Comment No. 18: Non-discrimination', (10 November 1989).
  'General Comment No. 31/80: The Nature of the General Legal Obligation Imposed on States Parties to the Covenant', (26 May 2004) CCPR/C/21/Rev.1/Add.13.
  Concluding Observations of the Human Rights Committee: United States of America, UN Doc. CCPR/C/USA/Q/3/CRP.4, para. 22 (2006).
IASC, 'IASC Operational Guidelines on the Protection of Persons in Situations of Natural Disasters' (The Brookings-Bern Project on Internal Displacement 2011).
ILC, Draft Articles on the Law of Treaties in Yearbook of the International Law Commission 1966 vol. II (1967) (A/CN.4/SER. A/1966/Add. 1).
  Fragmentation of International Law: Difficulties Arising from the Diversification and Expansion of International Law, UN Doc. A/CN.4/L.682 (April 13, 2006).

'Protection of Persons in the Event of Disasters: Comments and Observations received from Governments and International Organizations', (28 April 2016) UN Doc A/CN.4/696/Add.1.

'Draft Articles on the Protection of Persons in the Event of Disaster, Report of the Work of the 68th Session' (2 May–10 June and 4 July–12 August 2016) UN Doc A/71/10.

IPC Global Partners, 'Integrated Food Security Phase Classification Technical Manual Version 2.0: Evidence and Standards for Better Food Security Decisions' (Rome, FAO 2012).

IPCC, 'Summary for Policymakers'. In *Climate Change 2014: Impacts, Adaptation, and Vulnerability. Part A: Global and Sectoral Aspects. Contribution of Working Group II to the Fifth Assessment Report of the Intergovernmental Panel on Climate Change* (CUP 2014).

Manjoo R, 'Report of the Special Rapporteur on Violence against Women, Its Causes and Consequences', (2 May 2011) A/HRC/17/26.

OHCHR, *Protection of Internally Displaced Persons in Situations of Disaster: A Working Visit to Asia by the Representative of the United Nations Secretary General on the Human Rights of Internally Displaced Persons Walter Kälin* (Office of the UN High Commissioner for Human Rights 2005).

Sendai Framework on Disaster Risk Reduction 2015–2030, A/CONF.224/CRP.1.

The Limburg Principles on the Implementation of the International Covenant on Economic, Social and Cultural Rights (adopted 8 January 1987) UN Doc. E/CN.4/1987/17/.

UN Ad Hoc Committee on Refugees and Stateless Persons, 'Report of the Ad Hoc Committee on Statelessness and Related Persons' (Lake Success, New York, 16 January to 16 February 1950), 17 February 1950, E/1618; E/AC.35/5.

UN Economic and Social Council (ECOSOC), UN Economic and Social Council Resolution 672 (XXV): Establishment of the Executive Committee of the Programme of the United Nations High Commissioner for Refugees, 30 April 1958, E/RES/672 (XXV).

UNGA, Statute of the Office of the United Nations High Commissioner for Refugees (14 December 1950) A/RES/428(V).

'Climate Change and Its Possible Security Implications: Report of the Secretary-General' (11 September 2009) A/64/350.

UNGA, UNSC, 'Causes of Conflict and the Promotion of Durable Peace and Sustainable Development in Africa: Report of the Secretary General', (26 July 2016) A/71/211–S/2016/655.

UNHCR 'Report of the Working Group on Solutions and Protection to the Forty-Second Session of the Executive Committee of the High Commissioner's Programme (12 August 1991)', EC/SCP/64.

'Note on Burden and Standard of Proof in Refugee Claims', (16 December 1998).

'Interpreting Article 1 of the 1951 Convention Relating to the Status of Refugees', (April 2001).

'Environmental Migrants and Refugees', Refugees, 127 (2002).

'Guidelines on International Protection: "Internal Flight or Relocation Alternative" within the Context of Article 1A(2) of the 1951 Convention and/or 1967 Protocol Relating to the Status of Refugees', (23 July 2003) HCR/GIP/03/04.

'UNHCR's Refworld Case Law Collection User Guide, Status Determination and Protection Information Section (SDPIS)', (January 2009).

'Climate Change, the Environment, Natural Disasters and Human Displacement: A UNHCR Perspective', (14 August 2009).

'Forced Displacement in the Context of Climate Change: Challenges for States under International Law', Submission to the 6th Session of the Ad Hoc Working Group on Long-Term Cooperative Action under the Convention (20 May 2009), 9–10.

'Opening Statement by Mr. António Guterres, United Nations High Commissioner for Refugees, at the 60th Session of the Executive Committee of the High Commissioner's Programme (ExCom) (Geneva, 28 September 2009)' <http://www.unhcr.org/admin/hcspeeches/4ac314009/opening-statement-mr-antonio-guterres-united-nations-high-commissioner.html> accessed 16 April 2019.

António Guterres, 'Migration, Displacement and Planned Relocation', (31 December 2012) <http://www.unhcr.org/cgi-bin/texis/vtx/search?page=search&skip=342&docid=55535d6a9&query=water#_edn22> accessed 16 April 2019.

'Summary of Deliberations on Climate Change and Displacement', IJRL, 23 (2011), 561.

Handbook and Guidelines on Procedures and Criteria for Determining Refugee Status under the 1951 Convention and the 1967 Protocol Relating to the Status of Refugees (Reissued December 2011).

'Guidelines on Temporary Protection and Stay Arrangements', (February 2014) <http://www.refworld.org/docid/52fba2404.html> accessed 18 March 2018.

'States Parties to the 1951 Convention relating to the Status of Refugees and the 1967 Protocol', (April 2015) <http://www.unhcr.org/protect/PROTECTION/3b73b0d63.pdf> accessed 16 April 2019.

'Guidelines on International Protection No 12: Claims for Refugee Status related to Situations of Armed Conflict and Violence under Article 1A(2) of the 1951 Convention and/or 1967 Protocol Relating to the Status of Refugees and the Regional Refugee Definitions', (2 December 2016) HCR/GIP/16/12 <http://www.refworld.org/docid/583595ff4.html> accessed 16 April 2019.

UNHCR Executive Committee, 'Report of the Executive Committee of the Programme of the United Nations High Commissioner for Refugees: Conclusion on International Protection', (52nd Session, 1–5 October 2001) 5, A/56/12/Add.1 No. 90 (LII) – 2001.

UNHRC, 'Accelerating Efforts to Eliminate All Forms of Violence against Women: Ensuring Due Diligence in Prevention', (16 June 2010) A/HRC/14/L.9/Rev.1.

'The Right of Peoples to Peace: Progress Report Prepared by the Drafting Group of the Advisory Committee', (9 December 2011) A/HRC/AC/8/2

'Human Rights of Internally Displaced Persons', (29 June 2012) A/HRC/20/L.14.

'Report of the Independent Expert on the Situation of Human Rights in Haiti, Michel Forst, Addendum: Forced Returns of Haitians from Third States', (4 June 2012) A/HRC/20/35/Add.1.

'Report of the Special Rapporteur on Freedom of Religion or Belief, Heiner Bielefeldt', (29 December 2014) A/HRC/28/66; UNHRC, 'Annual Report of the Special Representative of the Secretary-General on Violence against Children' (30 December 2014) A/HRC/28/55.

'Report of the Special Rapporteur on the Rights of Indigenous Peoples, Victoria Tauli Corpuz', (6 August 2015) A/HRC/30/41.

'Report of the Special Rapporteur on the Human Rights of Internally Displaced Persons', (29 April 2016) A/HRC/32/35.

'Report of the Commission on Human Rights in South Sudan', (6 March 2017) A/HRC/34/63.

Living with Risk: A Global Review of Disaster Reduction Initiatives (Vol. 1, UNISDR 2004).

Terminology on Disaster Risk Reduction, 2009. Available at <http://www.unisdr.org/files/7817_UNISDRTerminologyEnglish.pdf> accessed 18 March 2018.

UNISDR/ESCAP, *Reducing Vulnerability and Exposure to Disasters: The Asia Pacific Disaster Report 2012* (UNISDR 2012).

UNOCHA, 'Philippines: Typhoon Haiyan Situation Report No. 28 (as of 31 December 2013)' <http://reliefweb.int/report/philippines/philippines-typhoon-haiyan-situation-report-no-28-31-december-2013> accessed 18 March 2018.

UNSC, '6587th meeting, (Resumption 1)', (20 July 2011) S/PV.6587.

'5663rd meeting', (17 April 2007) S/PV.5663.

News

Aisch G, A Pearce and K Yourish, 'Hurricane Irma Is One of the Strongest Storms in History', *New York Times* (9 September 2017) <https://www.nytimes.com/interactive/2017/09/09/us/hurricane-irma-records.html> accessed 16 April 2019.

BBC, 'Japan Earthquake: Explosion at Fukushima Nuclear Plant', *BBC* (12 March 2011) <http://www.bbc.com/news/world-asia-pacific-12720219> accessed 16 April 2019.

Chokshi N and M Astor, 'Hurricane Harvey: The Devastation and What Comes Next', *New York Times* (28 August 2017) <https://www.nytimes.com/2017/08/28/us/hurricane-harvey-texas.html> accessed 16 April 2019.

Elledge J, 'How the Fire at Grenfell Tower Exposed the Ugly Side of the Housing Boom', *New Statesman* (23 June 2017) <http://www.newstatesman.com/politics/uk/2017/06/how-fire-grenfell-tower-exposed-ugly-side-housing-boom> accessed 16 April 2019.

Gapper G, 'Grenfell: The Anatomy of a Housing Disaster', *Financial Times* (29 June 2017) <https://www.ft.com/content/5381b5d2-5c1c-11e7-9bc8-8055f264aa8b> accessed 16 April 2019.

Haygood W and AS Tyson, 'It Was as if All of Us Were Already Pronounced Dead', *Washington Post* (15 September 2005) <http://www.washingtonpost.com/wp-dyn/content/article/2005/09/14/AR2005091402655.html> accessed 16 April 2019.

McKibben B, 'Stop Talking Right Now about the Threat of Climate Change: It's Here; It's Happening', *The Guardian* (11 September 2017) <https://www.theguardian.com/commentisfree/2017/sep/11/threat-climate-change-hurricane-harvey-irma-droughts> accessed 16 April 2019.

Mundy S, 'Heavy Flooding Kills more than 1,000 in South Asia', *Financial Times* (30 August 2017) <https://www.ft.com/video/afc1b141-ad96-4084-875c-185d69e62a7c> accessed 16 April 2019.

Rushton C, 'Timeline: Hurricane Katrina and the Aftermath', *USA Today* (24 August 2015) <https://www.usatoday.com/story/news/nation/2015/08/24/timeline-hurricane-katrina-and-aftermath/32003013/> accessed 16 April 2019.

Watts J, 'Twin Megastorms Have Scientists Fearing This May Be the New Normal', *The Guardian* (6 September 2017) <https://www.theguardian.com/world/2017/sep/06/twin-megastorms-irma-harvey-scientists-fear-new-normal> accessed 16 April 2019.

INDEX

Aaldersberg and Others v Netherlands, 79, 134
AC (Tuvalu) case, 76, 80–81, 84, 86–87, 134
    NZIPT on, 78
Aceh, 49–50
Acosta case, 2, 96
AD (Tuvalu) case, 84–85
adversity, 1, 6, 47–48, 64–65, 82
    discrimination and, 68–69
    human agency in, 30
    indiscriminate, 48–54
    persecution and, 45, 92–93
    post-disaster, 63–64
    sea level rise and, 74
    state protection and, 66–67
    violence and, 63–64
AF (Kiribati) case, 54–55, 74, 76, 80–81, 83–84, 86–87, 134
    NZIPT in, 77, 79
Afghanistan, 32–33
African Court of Human and Peoples' Rights, 32
Anderson, Adrienne, 105, 131
anti-discrimination, 127–28. *See also* discrimination
Applicant A and Another, 52, 83, 89–90
    Dawson on, 2–4
asylum criteria, 89
At Risk: Natural Hazards, People's Vulnerability, and Disasters (Wisner), 18–19
Australia, 40. *See also* Federal Court of Australia
    persecution defined by, 70–71
    RSD in, 48
Australian Bureau of Meteorology, 139–40
AV (Nepal) case, 85

Bangladesh, 48
    cyclone in, 48–49
    state protection in, 68
Bankoff, G.
    on hazard paradigm, 14
    on vulnerability, 20
Bantu, 145–48
Barcelona Traction case, 112
Bayerisches Verwaltungsgericht Ansbach, 102
being persecuted. *See* persecution
BG Fiji case, 76

Botswana, 58
Briceño, Sálvano, 16
Bridge (Lord), 49
Budayeva case, 87–88
Burson, Bruce, 7–8
  on imminence, 80
  on RSD, 8

Canadian Immigration Appeals Board, 97–98
Cantor, D., 116–17
  on discrimination and persecution, 117
  on fundamental rights, 126
Cardoza-Fonseca case, 103–4, 141
Case 3719 II/58, 102
Castan, Melissa, 115
causation, 42
CEDAW. *See* Convention on the Elimination of All Forms of Discrimination against Women
CERD. *See* Convention on the Elimination of All Forms of Racial Discrimination
CESCR. *See* Committee on Economic, Social and Cultural Rights
Chan v Canada, 58, 141
Chen Shi Hai case, 41
Chetail, Vincent, 117
children, vulnerability of, 67
Christianity, 49
Cilacap, 141, 149
climate change
  anthropomorphising, 52
  in claims, 45–47
  defining, 10
  gender-based violence and, 64
  harm and, 74
  process of, 10
  RSAA on, 53
CMW. *See* Convention on Migrant Workers
Collins (Lord), 106–7
colonialism, 20, 28
Committee on Economic, Social and Cultural Rights (CESCR), 91, 114–15
  on discrimination, 114
communities
  defining, 18
  vulnerability and, 18
complementary protection, 79
Conference of Plenipotentiaries, 3
contributory cause approach, 144
  UNHCR on, 42
Convention on Enforced Disappearances, 112–13
Convention on Migrant Workers (CMW), 112–13
Convention on the Elimination of All Forms of Discrimination against Women (CEDAW), 91, 112–13
Convention on the Elimination of All Forms of Racial Discrimination (CERD), 112–13
Convention on the Rights of the Child (CRC), 67, 112–13
  Article 2, 113
  Article 3, 85
Convention Relating to the Status of Refugees, 1
conventional language, 121
Cox, Theodore N., 101
  on well-founded fear, 101–2

CRC. *See* Convention on the Rights of the Child
CRPD, 112–13
crucible approach, 121
customary international law, 91, 93–94, 112
   norms, 113
Cyclone Nargis, 55
   impact of, 55–56
   relief, 56
   RSD and, 56
cyclones, 139–40
   in Bangladesh, 48–49

Dalits, 18, 26
   vulnerability of, 26
Dawson (Judge), 4
   on *A and Another*, 2–4
   on natural disasters, 3
   on persecution, 3
denial, 130, 157–58
   violation and, 92–93
Desjardins (Judge), 58
Diplock (Lord), 103–4
disaster law, in EU, 13
disaster relief, 54
   Cyclone Nargis, 56
   discriminatory denial of, 57–63
   for famine, 58, 136
   humanitarian, 78, 83
discrimination, 28. *See also* non-discrimination norm; persecution
   adversity and, 68–69
   anti-discrimination, 127–28
   Cantor on, 117
   Committee on Economic, Social and Cultural Rights on, 114
   de facto, 114
   de jure, 114
   defining, 3, 22–28, 113, 122
   in disaster relief, 57–63
   ex-ante, 157–58
   Goodwin-Gill on, 128
   Hoffman on, 124
   human rights and, 123–24
   indirect, 114
   intention in, 42
   motivation for, 52, 71–72
   natural disasters and, 5–6
   persecution differentiated from, 3, 117, 119–28
   prevalence of, 23
   Refugee Convention on, 132–33
   social paradigm and, 22
   systemic, 114–15
   Twigg on, 25
   vulnerability and, 22–25
   Wisner on, 24
displacement, 25
   vulnerability to, 26–27

dissent, 54
    crackdowns on, 54–57, 157–58
DJ (Bantu – Not Generally At Risk) Somalia v Secretary of State for the Home Department, 146–47
dominant view (of disasters), 155
    Hewitt on, 12–13
    of RSD, 1–6
    Shacknove on, 5
Doymus case, 98–99
Draft Articles on the Protection of Persons in Situations of Disaster, 10, 78
    Article 3, 11
    Article 3(a), 12
    on natural disasters, 12
DS (Iran), 108, 123–24, 130
due diligence approach, 37
Duffy (Judge), 123–24
dynamic pressures, defining, 19

earthquakes, 69
    harm from, 73
    Lisbon earthquake of 1755, 13
    Nepal, 86
    in Peru, 20
ECHR. *See* European Convention on Human Rights
economic refugees, 48
Einarsen, Terje, 94, 121
Elias Zacarias case, 38
environmental damage
    intention and, 54
    the state as cause of, 54–55, 157–58
environmental degradation, 7–8
    human agency and, 78
    human rights and, 77–78
ethnic divisions, 24
    natural disasters and, 24–25
European Convention on Human Rights (ECHR)
    Article 2, 87–88
    Article 3, 60
European Court of Human Rights, 32
European Union, disaster law in, 13
event paradigm, 96–100
    persecution in, 97
    RSD under, 105–6

famine, 145–46
    relief, 58, 136
Famine Early Warning Service (FEWS), 147–48
    methodology of, 142
Farmer, Paul, on structural violence, 30–31
fascism, 101
fear. *See also* well-founded fear
    defining, 34–35
    Noll on, 34–35
    UNHCR on, 33–34
Federal Court of Australia, 39, 41–42
    on RSD, 66
Federal Emergency Management Agency, 152–53

Ferguson v Canada (Minister of Immigration and Citizenship), 63
FEWS. *See* Famine Early Warning Service
Fiji, 74
First Optional Protocol (to International Covenant on Civil and Political Rights), 79
Fletcher, Laurel E., 25–26
floods, 69
   control systems, 151
   in Pakistan, 57
   plains, 27–28, 74, 139
food aid, 61–62
food deprivation, 7
food insecurity, 142–48
   malnutrition, 67
   in Zimbabwe, 62
forecasting, 142
Foster, Michelle, 6, 43, 92–93, 105, 127–28, 131
   human rights approach of, 90
   on internal relocation, 44–45
   on motivation, 39
   on persecution, 35–36, 39
   RSD and, 92
   sustained and systemic formulation of, 99–100
Frank, Anne, 106, 108–9
Fugitive Offenders Act 1967, 103–4
Fukushima Daiichi Nuclear Plant, 70–71, 73–74
fundamental rights, Cantor on, 126

Gaillard, J. C., 24–25
Galtung, Johan, on structural violence, 28–30
gender-based violence, 63
   climate change and, 64
   natural disasters and, 64
Gilbert, G., 14
Gill, T., 26
Goff (Lord), 104
Goodwin-Gill, Guy S., 126–27
   on discrimination, 128
   on human rights, 127
   on motivation, 39
   on persecution, 35
   persecution defined by, 97
Goudron (Judge), 120
Grahl-Madsen, A., 35, 97, 102
   on persecution, 102, 119
Grenada, 63–64
Guatemala, 26–27
Guijt, J., 18
Guterres, António, 4

Hagi-Mohammed v Minister for Immigration and Multicultural Affairs, 58
Haiti, 30–31, 134–35
   structural violence in, 31
Hansch, Steven, 145–46
harassment, persecution and, 109
harm, 96
   climate change and, 74

from earthquakes, 73
intentional infliction of, 54–63, 69, 157–58
motivation to cause, 51–52
NZIPT on, 74
Refugee Review Tribunal on, 72–73
risks of, 98
Hathaway, James, 4–6, 43, 51, 57–58, 92–93, 106, 129–30
human rights approach of, 90
on IHRL, 90
on internal relocation, 44–45
on motivation, 39
on persecution, 35–36, 39, 61, 97–98
on Refugee Convention, 6
RSD and, 92
on state protection, 7
sustained and systemic formulation of, 99–100
VCLT interpreted by, 95
hazard events, 15
defining, 14–15
exposure to, 17
hazard paradigm and, 14
natural disasters and, 21–22
prediction of, 139
vulnerability and, 17
hazard paradigm, 12–15, 50, 155
Bankoff on, 14
defining, 12–13
hazard events and, 14
Hewitt on, 12–14
Hilhorst on, 13–14
influence of, 48
natural disasters and, 14–15
Refugee Review Tribunal on, 69
for RSD, 52
Hewitt, Kenneth
on dominant view, 12–13
on hazard paradigm, 12–14
High Court of Australia, 2, 39
Hilhorst, D., 13–14
Hindus, 57
HIV/AIDS epidemic, 30–31, 62
in Zimbabwe, 60
HJ and HT case, 106
Hoffman (Lord), 36, 40, 123, 125
on discrimination, 124
honour killing, 43
Hope (Lord), 4–5
Horvath case, 4–5, 83
HS (returning asylum seekers) Zimbabwe case, 59
human agency
in adversity, 30
environmental degradation and, 78
natural disasters and, 15, 155
persecution and, 78
structural violence and, 30
human rights, 35–36, 73, 98, 154. *See also* international human rights law

human rights (cont.)
    discrimination and, 123–24
    environmental degradation and, 77–78
    Foster on, 90
    Goodwin-Gill on, 127
    Hathaway on, 90
    international human rights law, 69–79
    in natural disasters, 31
    NZIPT on, 76–77
    RSD and, 7, 91, 118, 132–48
    UNHCR on, 36
    violation of, 87–88
humanitarian relief, 78, 83
Hurricane Harvey, 12
Hurricane Irma, 12
Hurricane Ivan, 63
Hurricane Katrina, 150–54

IARLJ, 96
ICCPR. *See* International Covenant on Civil and Political Rights
ICESCR. *See* International Covenant on Economic, Social and Cultural Rights
IFRC. *See* International Federation of Red Cross and Red Crescent Societies
IHRL. *See* international human rights law
Immigration Act of 2009 (New Zealand), Section 207, 84–85
imminence, 79, 87, 105–6
    Burson on, 80
    McAdam defining, 105
inaction, state protection and, 64
India, 26
Indian Ocean tsunami of 2004, 65
Indonesia, 49–50
inequalities, 18
    systemic, 115
Integrated Phase Classification (IPC), 62, 135–36
intention, 41. *See also* motivation
    in discrimination, 42
    environmental damage and, 54
    harm and, 54–63, 69, 157–58
    in nexus clause, 108
    relevance of, 120
Inter-Agency Standing Committee, 23
Inter-American Court of Human Rights, 32
Inter-Governmental Panel on Climate Change, 23–24
internal relocation, 149–50
    Foster on, 44–45
    Hathaway on, 44–45
    refugee status and, 44–45
International Association of Refugee Law Judges, 96
International Court of Justice, 32
International Covenant on Civil and Political Rights (ICCPR), 83, 112–13
    Article 6, 78, 80–82, 84, 134
    Article 7, 78, 80–82, 84, 134
    protection claims under, 78
International Covenant on Economic, Social and Cultural Rights (ICESCR), 91
    Article 2(2), 62–63
    Article 12(2)(b), 76–77

International Federation of Red Cross and Red Crescent Societies (IFRC), 23
international human rights law (IHRL), 69–79
　goals of, 91
　Hathaway on, 90–93
　non-discrimination norm in, 115–16
　obligations under, 130–31
　persecution and, 90–93
　the state role in, 92, 130–31
International Law Commission, 10, 78. *See also* Draft Articles on the Protection of Persons in Situations of Disaster
　on natural disasters, 10–11
international refugee law
　basic principles of, 33–34
　category 1 cases, 45, 48–54
　category 2 cases, 45, 54–63
　category 3 cases, 45
　non-discrimination norm in, 115–17
　peripheral cases, 47–48
　persecution in, 35–37
　scrutiny of, 49
　well-founded fear in, 33–34
interpretation
　of persecution, 98–99
　supplementary means of, 94
　of VCLT, 95
　of well-founded fear, 100–1
IPC. *See* Integrated Phase Classification
Islam case, 36, 40, 120, 123, 125, 130

Jakarta, 49
Japan, Refugee Review Tribunal on, 72
Joint Internally Displaced Person Profiling Service, 25
Joseph, Sarah, 115

K v Refugee Status Appeals Authority, 49
Kälin, Walter, 5, 149
Karanakaran case, 89
Katrina Canal Breaches Consolidated Litigation case, 154–55
Keith (Lord), 103–4
Kolmannskog, Vikram, 7

Lambert, Hélène, 98–99, 105, 131
Lauta, K. Cedervall, 13, 16
The Law of Refugee Status (Hathaway and Foster), 6, 92–93, 97–98
Linderfalk, Ulf, 121
Lisbon earthquake of 1755, 13
local risk index, 141

Mahler, C.
　on persecution, 39
　persecution defined by, 97
　on risk assessment, 102–3
Majid, M., 145–47
malnutrition, 67
McAdam, Jane, 131
　on imminence, 105
　on motivation, 39

McAdam, Jane (cont.)
   persecution defined by, 3–4, 35, 97
   on Refugee Convention, 48
   on risk assessment, 77–78
   on risk reduction, 73–74
   on RSD, 3
McDowell, S., 145–47
McHugh (Judge), 2, 89–90, 92, 104–5, 120–21, 141
   on persecution, 3, 119
meteorology, 139
methodology, 32–45, 73
   of FEWS, 142
   of NZIPT, 54, 86–87, 117
   of VCLT, 90, 92–95, 155–56
   of World Risk Index, 140–41
Michigan Guidelines, 42
Migration Act of 1958, 48
   Refugee Review Tribunal on, 73
   Section 5J(4), 42
Migration Act of 1998, Section 91, 52–53
Millett (Lord), 120
MIMIA v VFAY, 67
Minister for Immigration and Multicultural Affairs v Respondents S152/2003, 92
Mohammed Motahir Ali v. Minister of Immigration, Local Government, and Ethnic Affairs, 48
motivation, 68, 70–87
   discriminatory, 52, 71–72
   Foster on, 39
   Goodwin-Gill on, 39
   harm in, 51–52
   Hathaway on, 39
   McAdam on, 39
   for persecution, 38–42
   of the state, 66–67
Muslims, 57
Myanmar, 55
   activism in, 55

Nansen Initiative on Disaster-Induced Cross-Border Displacement, 23–24
nationality, 44
   RSD and, 101
natural disasters, 4–5, 140–41
   in claims, 45–47
   Dawson on, 3
   defining, 1, 10, 21–22
   discrimination and, 5–6
   Draft Articles on the Protection of Persons in Situations of Disaster on, 12
   ethnic divisions and, 24–25
   future scenarios, 133–36
   gender-based violence and, 64
   hazard events and, 21–22
   hazard paradigm and, 14–15
   human agency and, 15, 155
   human rights in, 31
   International Law Commission on, 10–11
   NZIPT on, 81
   ongoing situations, 133–34

poverty and, 25–26
Refugee Convention and, 5–6, 70
Refugee Review Tribunal on, 71
refugee status and, 31
Refugee Status Review Committee on, 49
Shacknove on, 6
slow-onset, 20, 142
social paradigm and, 14–15
state protection and, 87
sudden-onset, 20
triggering, 14
violence and, 63
Nazis, 101, 106
Nepal earthquake, 86
New Zealand, 40
  RSD in, 76
New Zealand High Court, 82–83
New Zealand Immigration and Protection Tribunal (NZIPT), 49–52, 54, 100, 117–18
  in *AC (Tuvalu)* case, 78
  in *AF (Kiribati)*, 77, 79
  on harm, 74
  on human rights, 76–77
  intention of, 54
  jurisprudence, 87
  methodology of, 54, 86–87, 117
  on natural disasters, 81
  on persecution, 123
  Refugee Convention and, 83
  on risk assessment, 80–81
  on RSD, 58–59
  on sea level rise, 75–76
  on state protection, 74
nexus clause, 122–24, 128–29, 143–48
  intention in, 108
  non-discrimination norm in, 117–18
Nigeria, vulnerability in, 19
Noll, Gregor, 34–35
non-discrimination norm, 112–15
  in IHRL, 115–16
  in international refugee law, 115–17
  new theoretical model and, 116
  in nexus clause, 117–18
  in persecution, 118–21, 124–25
non-refoulement, 78, 80, 91, 105–6
NZIPT. *See* New Zealand Immigration and Protection Tribunal

Oliver-Smith, Anthony, 15–16
  on PAR model, 20
Omar Mohamud Mohamud v MIMA, 66
Omoruyi case, 127–28
Ontario Federal Court, 64
oppression, 127
  persecution differentiated from, 92–93

Padang, 141
Pakistan, floods in, 57

PAR model. *See* pressure and release model
patriarchy, 28
Paul, Bimal Kanti, 136–39
Peacock, Walter Gillis, 18
Perampalam case, 149–50
Perry, R., 12
persecution, 2–3, 40, 43. *See also* intention; motivation
   actors identified in, 52
   adversity and, 45, 92–93
   Australian statutory definition of, 70–71
   being persecuted, 96, 118–22, 129–31
   Cantor on, 117
   Dawson on, 3
   defining, 4, 35–37, 87–88, 96, 98, 109, 121, 128–31
   discrimination differentiated from, 3, 117, 119–28
   in event paradigm, 97
   Foster on, 35–36, 39
   Goodwin-Gill on, 35
   Grahl-Madsen on, 102, 119
   harassment and, 109
   Hathaway on, 35–36, 39, 61, 97–98
   human agency and, 78
   international human rights law and, 90–93
   in international refugee law, 35–37
   interpretations of, 98–99
   liberal school of, 97
   Mahler on, 39
   McAdam on, 3–4, 35
   McHugh on, 3, 119
   motivation for, 38–42
   non-discrimination norm in, 118–21, 124–25
   NZIPT on, 123
   oppression differentiated from, 92–93
   Qualification Directive on, 108
   reasons for, 37–38
   restrictive school of, 97
   risks of, 98
   temporal scope of, 107, 122
   UNHCR on, 61
   Zimmerman on, 39
persistency, requirements for, 98–99
personal scope of refugee definition, 68, 87, 95, 155
   defining, 130
Peru, 20
Pobjoy, J., 106
political beliefs, 44
political ecology, 15–16, 21–22
poverty, natural disasters and, 25–26
precautionary planning, 15
predicament approach, 41–42
   defense of, 108
   judicial development of, 108
   of RSAA, 89–90
   in state protection failures, 66–67
   to well-founded fear, 107–11

pressure and release (PAR) model, 22
    Oliver-Smith on, 20
    unsafe conditions aspect of, 20
    of vulnerability, 18–21
    Wisner on, 20–21
Priestly (Judge), 83
private sphere, 130
proof, standard of, 102–3
    well-founded, 148–49
protection (state), 35–36, 121
    adversity and, 66–67
    in Bangladesh, 68
    ex-ante failures of, 69–79
    ex-post failures of, 63–68
    failures of, 37, 63–87, 157–58
    Hathaway on, 7
    inaction and, 64
    natural disasters and, 87
    NZIPT on, 74
    predicament approach and, 66–67
    RSAA on, 65
protest, 57
public order, 157–58
public sphere, 130

Qualification Directive, 61, 96
    Article 9(1), 99
    Article 9(3), 118
    persecution defined by, 108

R v Governor of Pentonville Prison ex Parte Fernandez, 103–4
R v Secretary of State for Home Department, Ex p Bugdaycay, 49
race, 44
racism, 28, 113
Ragsdale, A. K., 18
Ram v MIEA and Anor, 52, 120
real risk, 110
reasonable grounds, 103
reasonable person test, 102
Reewin, 145, 147–48
Refugee Appeal No. 11/91, 76
Refugee Appeal No. 70074, 117
Refugee Appeal No. 70959/98, 65
Refugee Appeal No. 70965, 64–65
Refugee Appeal No. 72186/2000, 53
Refugee Appeal No. 72635/01, 41, 43
Refugee Appeal No. 74655/03, 76, 89–90
Refugee Appeal No. 74665/03, 110
Refugee Appeal No. 76237, 60
Refugee Appeal No. 76374, 55, 77–78
Refugee Convention, 36–37, 50–51, 82–83, 93, 129, 143
    applicability of, 77
    Article 1A(2), 6, 33–34, 38–39, 44–45, 72–73, 77, 83, 107, 109, 116, 122, 125–26, 129, 150
    Article 33, 35
    on discrimination, 132–33
    goals of, 94

Refugee Convention (cont.)
    Hathaway on, 6
    limits of, 4
    McAdam on, 48
    natural disasters and, 5–6, 70
    NZIPT on, 83
    Refugee Review Tribunal on, 68, 72–73
    RSAA on, 53
    on RSD, 132–33
refugee law. *See* international refugee law
Refugee Review Tribunal (of Australia), 65–67, 69
    on harm, 72–73
    on hazard paradigm, 69
    on Japan, 72
    on Migration Act of 1958, 73
    on natural disasters, 71
    on Refugee Convention, 68, 72–73
    on religion, 71
    on RSD, 69
Refugee Status Appeals Authority of New Zealand (RSAA), 41, 43, 55–56, 107–8
    on climate change, 53
    predicament approach of, 89–90
    on Refugee Convention, 53
    on state protection, 65
refugee status determination (RSD), 1, 84, 156. *See also* methodology; *specific topics*
    in Australia, 48
    Burson on, 8
    Cyclone Nargis and, 56
    defining, 45
    dominant view of, 1–6
    under event paradigm, 105–6
    Federal Court of Australia on, 66
    Foster on, 92
    Hathaway on, 92
    hazard paradigm for, 52
    human rights and, 7, 91, 118, 132–48
    internal relocation and, 44–45
    McAdam on, 3
    nationality and, 101
    in New Zealand, 76
    New Zealand Immigration and Protection Tribunal on, 58–59
    Refugee Convention on, 132–33
    Refugee Review Tribunal on, 69
    risk assessment in, 53–54
    Shacknove on, 6
    test for, 132
    unitary character of, 89–90
    vulnerability and, 66–67
Refugee Status Review Committee, 49
    on natural disasters, 49
ReliefWeb, 135
religion, 44
    Refugee Review Tribunal on, 71
risk
    exposure, 22
    harm and, 98

real, 110
reduction, 73–74
risk assessment, 108–10, 130, 139–40
    Mahler on, 102–3
    McAdam on, 77–78
    NZIPT on, 80–81
    in RSD, 53–54
    well-founded fear as, 100–7
    Zimmerman on, 102–3
RN (Returnees) Zimbabwe CG case, 58–62
Robinson, Jacob, 3
Rodger (Lord), 106
Rodríguez, Havidán, 22
Routledge Handbook of Hazards and Disaster Risk Reduction, 18–19
RRT Case No. 0901657, 49
RRT Case No. 0903555, 56
RRT Case No. 1001325, 57
RRT Case No. 1104064, 57
RRT Case No. 1200203, 69–70
RRT Case No. 060926579, 57
RRT Case No. 071295385, 65
RRT Case No. N93/00894, 68
RRT Case No. N96/10806, 50–51
RS and Others case, 60
RSAA. *See* Refugee Status Appeals Authority of New Zealand
RSD. *See* refugee status determination

S v Chief Executive of the Department of Labour and Anor, 49
salination, 74
    sea level rise and, 75–76
Scalia (Justice), 38–39
Schrepfer, Nina, 5
sea level rise, 74, 81
    adversity and, 74
    NZIPT on, 75–76
    salination and, 75–76
Sedley (Justice), 89, 120–21, 148–49
Serje, J., 24
Shacknove, Andrew, 5
    on dominant view, 5
    on natural disasters, 6
    on refugee status, 6
Shah, M. K., 18
sharecropping, 146–47
Sivakumaran case, 103–4, 141
slow violence, 31
small-scale disasters, 11–12
social groups, 44
social paradigm, 8, 14–16
    Cedervall Lauta on, 16
    discrimination and, 22
    natural disasters and, 14–15
    vulnerability in, 16–18
    Wisner on, 26–27
Söderbergh, Carl, 26
Somalia, 136–40, 142, 145–48

Sri Lanka, 65–68
the state
   environmental damage caused by, 54–55, 157–58
   IHRL and role of, 92, 130–31
   motivation of, 66–67
   violence from, 57
state practice, 32–33
state protection. *See* protection (state)
Stevens (Justice), 38–39
Steyn (Lord), 125
Storey, Hugo, 98–99
Stover, Eric, 25–26
structural violence, 28–31, 108, 110
   defining, 29
   Farmer on, 30–31
   Galtung on, 28–30
   in Haiti, 31
   human agency and, 30
   root causes of, 31
   slow, 31
   temporal scope of, 31
   UN Special Rapporteur on the Human Rights of Internally Displaced Persons on, 29
super-majority principle, 113, 115–16
Supreme Court (Canada), 2
Supreme Court (New Zealand), 84
Supreme Court (UK), 2, 106
Supreme Court (US), 38
surrogacy, principle of, 4–5, 44–45, 94
Suu Kyi, Aung San, 56

Taliban, 32–33, 57, 67
Tamil, 65
technocratic intervention, 13
Teitiota, Ioane, 74
Teitiota v The Chief Executive of the Ministry of Business Innovation and Employment, 82–83
temporal scope, 68, 109–10
   of persecution, 107, 122
   of structural violence, 31
Togo, 56–57
treatment, 84
   defining, 78
tsunamis, 65
   Indian Ocean tsunami of 2004, 65
Tuvalu, 50–51, 53, 75
   government of, 79
Twigg, J., 20–21
   on discrimination, 25

UDHR. *See* Universal Declaration on Human Rights
Ullah v Special Adjudicator, 61
UN Framework Convention on Climate Change (UNFCCC), 10
UN General Assembly, 101
UN High Commissioner for Refugees (UNHCR), 118–19
   on contributory cause approach, 42
   on fear, 33–34

on human rights, 36
on persecution, 61
UN Human Rights Committee, 25, 32, 134
UN Office for Disaster Risk Reduction (UNDRR), 11–12, 16
   on vulnerability, 17
UN Office for the Coordination of Humanitarian Affairs (UNOCHA), 135, 145
UN Special Rapporteur on the Human Rights of Internally Displaced Persons, 25
   on structural violence, 29
   on vulnerability, 26–27
UNDRR. *See* UN Office for Disaster Risk Reduction
UNFCCC. *See* UN Framework Convention on Climate Change
UNHCR. *See* UN High Commissioner for Refugees
UNISDR. *See* UN Office for Disaster Risk Reduction
United Kingdom, 40
   protection obligations of, 59
Universal Declaration on Human Rights (UDHR), 90, 94
   Article 2, 125
   on persecution, 97
UNOCHA. *See* UN Office for the Coordination of Humanitarian Affairs
unsafe conditions, in PAR model, 20

VCLT. *See* Vienna Convention on the Law of Treaties
Venezuela, 24
Vernant, Jacques, 97
Vienna Convention on the Law of Treaties (VCLT), 2, 90–91
   Article 31, 90, 93–95, 109–10, 115–16, 121, 130–31, 155–56
   Article 31(1), 93–94
   Article 31(2), 93–94
   Article 31(3), 94
   Article 31(3)(c), 113, 115–16
   Article 31(4), 109–10
   Article 32, 90, 93–95, 130–31, 155–56
   Article 33, 90, 93–95, 130–31, 155–56
   Hathaway on, 95
   methodology prescribed by, 90, 92–95, 155–56
violation, denial and, 92–93
violence. *See also* structural violence
   adversity and, 63–64
   gender-based, 63–64
   natural disasters and, 63
   state, 57
vulnerability, 152
   Bankoff on, 20
   causes of, 67
   of children, 67
   communities and, 18
   of Dalits, 26
   defining, 19
   discrimination and, 22–25
   to displacement, 26–27
   hazard events and, 17
   in Nigeria, 19
   PAR model of, 18–21
   root causes of, 19
   RSD and, 66–67
   in social paradigm, 16–18

vulnerability (cont.)
    UN Special Rapporteur on the Human Rights of Internally Displaced Persons on, 26–27
    UNDRR, 17

de Waal, Alex, 145–46
Ward case, 83
Weinstein, Harvey M., 25–26
Weis, Paul, 35, 97
well-founded fear, 37, 92, 155
    circumstances not reflecting, 150–55
    Cox on, 101–2
    establishing, 52
    in international refugee law, 33–34
    interpretation of, 100–1
    predicament approach to, 107–11
    as risk assessment, 100–7
    subjectivity of, 102–3
Whitlam (Judge), 48
Wisner, Ben, 18–19
    on discrimination, 24
    on PAR model, 20–21
    on social paradigm, 26–27
World Disasters Report, 23
World Risk Index, 140, 143
    methodology of, 140–41

YC v Holder, 57

Zimbabwe, 59
    food insecurity in, 62
    HIV/AIDS epidemic in, 60
Zimmerman, A., 94
    on persecution, 39, 97
    on risk assessment, 102–3
Zink, Karl Friedrich, 97